Fuel

Environmental Cultures Series

Series Editors:
Greg Garrard, University of British Columbia, Canada
Richard Kerridge, Bath Spa University

Editorial Board:
Frances Bellarsi, Université Libre de Bruxelles, Belgium
Mandy Bloomfield, Plymouth University, UK
Lily Chen, Shanghai Normal University, China
Christa Grewe-Volpp, University of Mannheim, Germany
Stephanie LeMenager, University of Oregon, USA
Timothy Morton, Rice University, USA
Pablo Mukherjee, University of Warwick, UK

Bloomsbury's **Environmental Cultures** Series makes available to students
and scholars at all levels the latest cutting-edge research on the diverse ways
in which culture has responded to the age of environmental crisis. Publishing
ambitious and innovative literary ecocriticism that crosses disciplines, national
boundaries, and media, books in the series explore and test the challenges of
ecocriticism to conventional forms of cultural study.

Titles available:
Bodies of Water, Astrida Neimanis
Cities and Wetlands, Rod Giblett
Civil Rights and the Environment in African-American Literature, 1895–1941,
John Claborn
Climate Crisis and the 21st-Century British Novel, Astrid Bracke
Ecocriticism and Italy, Serenella Iovino
Fuel, Heidi C. M. Scott
Literature as Cultural Ecology, Hubert Zapf
Nerd Ecology, Anthony Lioi
The New Nature Writing, Jos Smith
The New Poetics of Climate Change, Matthew Griffiths
This Contentious Storm, Jennifer Mae Hamilton

Forthcoming Titles:
Anthropocene Romanticism, Kate Rigby
Cognitive Ecopoetics, Sharon Lattig
Colonialism, Culture, Whales, Graham Huggan
Eco-Digital Art, Lisa FitzGerald

Fuel

An Ecocritical History

Heidi C. M. Scott

BLOOMSBURY ACADEMIC
LONDON • NEW YORK • OXFORD • NEW DELHI • SYDNEY

BLOOMSBURY ACADEMIC
Bloomsbury Publishing Plc
50 Bedford Square, London, WC1B 3DP, UK

BLOOMSBURY, BLOOMSBURY ACADEMIC and the Diana logo are trademarks of
Bloomsbury Publishing Plc

First published in Great Britain 2018

A catalogue record for this book is available from the British Library.

Library of Congress Cataloging-in-Publication Data
Names: Scott, Heidi C. M. (Heidi Cathryn Molly), 1978– author.
Title: Fuel: an ecocritical history / by Heidi C.M. Scott.
Description: London ; New York : Bloomsbury Academic, 2018. |
Series: Environmental cultures series | Includes bibliographical references and index.
Identifiers: LCCN 2017060189 (print) | LCCN 2017060449 (ebook) | ISBN 9781350053991 (ePDF) |
ISBN 9781350054004 (ePUB) | ISBN 9781350053984 (hardback : alk. paper) |
Subjects: LCSH: Fuel–Popular works. | Power resources–Popular works.
Classification: LCC TP318 (ebook) | LCC TP318 .S375 2018 (print) | DDC 662.6–dc23
LC record available at https://lccn.loc.gov/2017060189

ISBN: HB: 978-1-3500-5398-4
ePDF: 978-1-3500-5399-1
eBook: 978-1-3500-5400-4

Series: Environmental Cultures

Typeset by Newgen KnowledgeWorks Pvt. Ltd., Chennai, India
Printed and bound in Great Britain

To find out more about our authors and books visit www.bloomsbury.com
and sign up for our newsletters.

For my parents, diesel in intellect and solar in support:
Bonnie Kime Scott
and
Thomas Russell Scott
(1944–2017)

Contents

Figures

Acknowledgments

I am grateful for the assistance of many friends and colleagues on the journey to press. The Association for the Study of Literature and the Environment (ASLE) community provided early suggestions for primary sources, and the preexisting work of its scholars was a foundation on which to build greater perspective into the energy humanities. Florida International University provided essential funds to enable the early research stages of the manuscript, and assistants Dina Lopez and Jennifer Kenneally combed through online archives armed with keywords. The University of Maryland continues to support my teaching and scholarship. Series editors Greg Garrard and Richard Kerridge generously advised and improved the manuscript through revisions and the peer review process. Readers Kathleen McCormick, Joseph Byrne, Bonnie Kime Scott, Colleen Flaherty, and Nina Gayleard provided insightful comments that helped flag my own energetic assumptions. Bloomsbury editors David Avital and Clara Herberg were responsive and personable amid the final flurry.

Thanks to you all.

Introduction

I New York City to Albany

A magic space-travel chair awaits you. It moves at 550 miles per hour, about two hundred times faster than you could move your own body over long distances. At thirty thousand feet, the chair flies above the clouds and affords you a bird's-eye view of earth. You traverse 150 miles nearly due north, lifting clean out of the snags of America's largest city. It safely delivers you within an hour to your far-flung destination, the state capital of Albany. Would you like African java, Indian tea, or one of a dozen corn syrup drinks while hurtling through thin air? There are fifty magic chairs on an airplane, in this case a small jet, and at this moment about ten thousand jets are in flight.

There is an alternative magic chair. This one averages sixty miles per hour over the ground, and takes about 2.5 hours to cross New York State. It is amazingly affordable and is continuously, redundantly available whenever you choose to leave the city. There is minimal danger to your body, though more than in the flying chair. Barring traffic, you will probably never need to slacken your superhuman pace during the 150-mile journey. If you do stop, every mass-produced delicacy known to modern society will be offered at multiple competing outlets.

A third chair also moves around sixty miles per hour, but this one rocks on rails and conforms to a time table. One advantage over the second chair is that it is completely self-navigating. This chair is propelled by a different refinement of the same stupendously powerful fuel that pushes chairs one and two: jet fuel, gasoline, and diesel, respectively. Oil. Slow-cooked plankton millions of years old. All three seats are departing right now.

Our near ancestor occupies a fourth chair on a boat. The 1807 river cruise from New York to Albany took about thirty hours, averaging five miles per hour. The *North River* steamboat was the first to navigate that length of the Hudson as

Figure 1.1 The *Clermont*, a replica of the *North River* steamboat, operates on the Hudson in 1909. The original voyage to Albany took place in 1807. Image courtesy of Wikimedia Commons via Library of Congress. Public domain.

a viable commercial ferry. Both of her paddle wheels turned by coal-fired steam, and she had supplemental masts with sails and rigging, a chimera run by fire and wind—a smoke-snorting dragon with pterodactyl wings. On this trip, though, the sails stayed furled. The *North River* burned ancient sunlight, materialized into carbon chains by plants, and then compressed by geologic plates. Coal. She quickly passed sailing sloops tacking windward, taking a straight course north toward her destiny. Her only trace past the first bend in the river was the gray fuzz of carbon fixed from Carboniferous-era air. After three hundred million years, the carbon is free again in the Anthropocene. We are liberated, too, in a way, from slow, disperse fuels formed from sun, wind, tides, wood, grass, and food, to the much denser fossil fuels found in geologic cemeteries. These load-liberating fossil fuels also liberate greenhouse gasses, causing climate change, civilization's leading challenge into the twenty-first century.

Now the nature of the chair changes. We are moving backward through time, and must exchange our lavish fossil-fuel-flown thrones for bare leather saddles. It's the late seventeenth century, and your saddle pack is stuffed with English silver to trade with Dutch trappers for fur. Keeping your fourth seat requires horsemanship, energy, endurance, and knowledge of the landscape. You may seek recourse to your pistol or fists along the way, depending on the motley company that loiters on the cliffs of the Hudson River. Your horse will require rest and feed, and you will need a new animal altogether if you wish to travel continuously to finish the journey in four days. If all goes well, you will have barely used linens in the bed at the inn, and also vittles and ale to wash down the dust. If not, you may have a leaf tuft beneath a stout oak and a saddle bag picnic. Nighttime brings a prayer that the local wolves and panthers do not visit your camp to kill and devour your chair.

The fifth seat tacks against the wind and tide on the ever-changing Hudson. You rarely sit down, and share a bunk bed in shifts. Your calloused hands follow shouted orders in concert with a dozen other laboring sailors as you stalk the fitful engine of the wind: elemental or primary energy. If you are Henry Hudson sailing in 1609, you hope that you will find China upstream, but once you traverse 150 miles in ten days (an average of 0.63 lateral miles per hour), you elect to return to Holland. Over the next few hundred years, your turn-around spot will become Albany, and sailing ships bearing goods from the interior will crowd the waters. China never materializes upstream, but the wind is free and abundant and sometimes blows the right way, allowing for the exchange of the rich natural resources of the interior, like timber and hides, for traded goods from Europe and points oriental.

The sixth is no chair at all, but a footpath. Before the colonial period, Native Americans used their muscles, powered by digested plants and animals, to traverse long distances over the wild terrain of the future Empire State. It would have taken at least a week to walk from the mouth of the Hudson upstream 150 miles, for the few who chose to wander so far from home. Such paths would have been taken for trade. Humans have been striking out on foot for our entire two hundred thousand–year history, and before that we did it on slightly different legs, most likely those of the ancestor species *Homo heidelbergensis*. Whether horse or human, these journeys were fueled by biomass—food like grains, fruits, vegetables, and some meat. Within a year this biomass fuel would be replenished by the sun.

II Fuel history

Of course, we use fuel for many tasks besides transportation. Besides the fact that our bodies are furnaces that require the fuel of regular meals, in modern society a great proportion of fuel also goes into industrial production, construction and maintenance of infrastructure, heating and cooling, and the exchange of information and entertainment. The contrast among these six forms of transportation shows how radically and rapidly our lives shift due to changes in the fuel sources propelling society. In the last 250 years, the transition of energy sources from biomass and primary energy to fossil fuels—coal, oil, and natural gas—fundamentally affected the human experience of self, time, and place. By unleashing the astonishing power of fossil fuels using the steam engine, coal introduced more changes to life, environment, and society than the world had seen in the previous twelve thousand years of the agricultural era. Europe first used fossil fuels on a large scale in stationary industry in the 1700s, followed by locomotives and steamships in the 1800s. By the early 1900s, petroleum took some of the burden off coal: it again revolutionized movement, now by personal cars, and was useful as a stock material for an explosion of consumer goods: paints, plastics, and chemicals. This shift in bioenergetics propelled us beyond the ancient constraints of a biomass economy into a realm of modern industrial superpower built upon prehistoric carbon. Petroleum is an acme: it is the only fuel dense enough to make itself fly, when powering a prop or jet engine.

We may not realize the fantastically larger amounts of fuel we use in developed nations compared to our predecessors in preindustrial societies, both the agricultural societies of the last twelve thousand years and the hunter-gatherer societies that subsisted for hundreds of millennia before that. Energy consumption correlates with cultural evolution. For 95 percent of our existence as a species, we have lived as hunter-gatherers, mainly on the calories needed to keep our bodies fueled, plus a bit of biomass for fires. Agriculture tripled that rate by introducing the labor of domestic animals needing fodder, the calories provided by their milk and meat, and the grain harvests enabled by their labor. By the early modern period, agriculture was accentuated by widespread metallurgy and glassmaking that required furnaces fueled by charcoal, a processed form of biomass. Once we figured out how to intensify industrial fires with coke derived from coal, rather than charcoal, the stage was set for mass industrialization and, with it, mass consumerism. This was the single largest leap in energy consumption in history, made visible in the coal-gray skies hulking above nineteenth-century cities.

By 1900, people in industrial culture were using more than twenty-two times the number of daily calories as ancestral humans. Between 1850 and 1970, Americans witnessed a 150-fold increase in energy consumption despite improvements in efficiencies (Johnson 5). Consuming ever-greater amounts of energy was an invisible buoy of the American dream, in which each generation could observe gains in energy use beyond their parents' time. These gains in energy use are commonly coded as "successories": higher incomes that propagated larger houses, yards, cars, wardrobes, appliances, vacations, and toys.

We live in an era in which fuel is ubiquitous in our lives, but our experience of it is largely immaterial. Fuel is not invisible: the landscape of the average early twenty-first-century city in a developed country is filled with highways and gas stations, power lines and charging stations. However, this abundant energy network spares the energy consumer most of the physical, mental, and social effort that was once required to secure enough energy to survive. The average consumer's body no longer performs the work of harvesting, preparing, and releasing energy contained in fuel, as the preindustrial forest dweller felled, split, cured, and burned the wood that permitted his community to endure winter and cook food that would be inedible in raw form. Today, and for several generations now, the idea of energy has been growing increasingly abstract. Abundance and scarcity are not entwined with a physically observable extent of forest cover. They have become tabulations of oil and coal reserves, solar and wind installations, presented as numbers in newspapers, or, more likely, shine on websites in which the relationship between the actual energy supply and the physical experience of using that energy—the coal burning to illuminate the laptop—are fully abstracted from one another. Similarly, the energy network spares us the work of thinking about, let alone performing, the fuel tasks necessary to cook food. The steam emanating from a bowl of microwaved oatmeal is energy stored from the photosynthesis of Carboniferous-era trees, cured by deep time, compression, and heat into a rock of coal. This energy is liberated by a minerals corporation from a mine in the next state or around the world, pulverized to powder, and burned in a diabolically hot fire that heats water to steam and turns turbines in a nearby power plant. Finally, the grid transmits that energy as electricity into our home, our microwave, and our oatmeal.

We look to the sky and see blue or white, not the gray smog that oppressed early industrial cities in which coal smoke blew from every chimney. As our skies display much less of the visible malaise of industrial-level energy consumption than our recent forbears endured, our fingernails bear no grime from the hard work of energy extraction. (More than ever, machines do the work that laborers once performed to harvest fuel like coal, and clean fingernails across the world tap at keyboards.) Locales that do suffer from oppressive, palpable pollution tend to be centers of industrial production, like the Chinese city of Baoding, or centers of individual oil consumption, like the highways in the bowl of greater Los Angeles. These exceptions that are in parallel between energy and experience only make the general paradox more striking: we consume more energy per capita than any humans ever have before, and we invest minimum cognitive, physical, and temporal effort, and experience less discomfort than any of the fuel-burning *Homo sapiens* gathered around open fires for the last two hundred thousand years. And yet (here's the rub), the collective result of our energy binge is the colossal problem of climate change that has only come to the full social and political consciousness in the last generation. With fuel, we have a lighter physical burden and a heavier moral burden than any of our ancestors.

In using this general *we*, I am of course eliding the differences between socioeconomic levels that make exposure to the negative aspects of energy use a varied experience. Poor people in the Houston area are more likely to suffer from the air and water pollutions inherent to petroleum refineries than their wealthy counterparts, who can afford to buy real estate less tainted by the local industry (those very people who benefit most from the industry's profits). This story of socio-environmental geography is nothing new. Queen Elizabeth removed to Windsor Castle to avoid the coal smoke endemic to early modern London, which defaced building facades and choked out trees long before the modern industrial era (Hiltner 316); Frederich Engels observed laborers' hovels in the industrial labyrinths of mid-nineteenth century Manchester; Charles Dickens created Coketown as a study of urban laborers' oppression juxtaposed to the pastoral suburbs of the wealthy; Upton Sinclair updated these visions for the petroleum industry of early twentieth-century southern California, which displaced subsistence homesteaders and pulled the culture into the thrall of oil speculation and get-rich-quick schemes; today, environmental justice focused especially on racial and income disparities is a major subfield in the social sciences and humanities, public policy, and environmental regulation.

Still, for most people in developed nations, energy is secured by an online electricity payment and a few swings into the gas station and the grocery store. In return, we enjoy a practically limitless supply. Of course, the individual American experience of infinite, cheap, high-quality energy (at this writing, the average gas price in several US states is less than $2.50/gallon) is an illusion supported by a global network of corporate colonialism, industry-driven regulation standards, and federal subsidies. We live in a strange era in which fossil fuels signify both abundance and the threat of scarcity, enjoyment and fear, exuberance and melancholy: this is our fuel ontology.

III Defining *Fuel*

At this point, a definition and justification of terms is in order. To the casual reader, in the preceding paragraphs I have used the words "energy" and "fuel" interchangeably. But they actually preserve an important distinction. *Energy* denotes both a literal, material entity known as fuel, but it may also be used to obfuscate between fuel and metaphorical concepts such as the "energy" of intellect, ingenuity, collaboration. Allen MacDuffie's study of energetics and thermodynamics in Victorian England wisely dismantles the rhetoric of a Shell Corporation advertisement that slyly substitutes "human energy" for diminishing and increasingly problematic supplies of fuel (1). As MacDuffie observes, this rhetorical sleight of hand cons the viewer into believing that the abstract, disembodied energy that emanates from human cognition and cooperation is equivalent to, and even greater than, the "limits brute matter would impose" (1). Succinctly, there is no energy crisis. We have a *fuel* crisis (2).

This book is about brute matter. I titled it *Fuel* because I mean to indicate the physical substance used in specific eras of history to provide energy. Wherever possible, I wish to disentangle the material denotation of fuel from its metaphorical lives as energy, which are often deployed without much awareness of materiality. (The *Oxford English Dictionary* records both material and abstract uses of the word "energy" first dating to the early 1600s; material "fuel" dates much earlier to 1398.) Laurie Shannon has identified a problem with the elision of fuel materiality: "Western culture has transitioned to forms of energy whose origins are opaque to ordinary perception, whose material workings are comprehended only by specialists, and whose business operations are shielded and securitized. One result seems clear. Literally visceral knowledge of where energy comes from, or what energy is, has been substantially extinguished" (312). Unlike some enlightening studies on the metaphorical lives of energy in specific cultures, such as MacDuffie's focus on the idea of thermodynamics in early industrial England, this book is focused on the *material cultures of fuel* within the three distinct bodies of biomass, fossil, and new renewable energy.

Writers have for centuries taken account of how our human condition directly relates to fuel sources. Usually in literature, though, the relationship is obscured and implicit. It is the aim of this book to make it overt and explicit. There is no better source material to study the history of fuel than literature. The technology and sociology of fuel is already well-represented by scholars like Vaclav Smil, but literature provides source texts that are at once historical, rhetorical, and emotional, thus opening ecocritique to new perspectives on how energy histories inform the nature of our existence—as humans, and as twenty-first century consumers in the Anthropocene. This book uses mostly Anglophone literature—with a few excursions into translated works from Europe—to explore the fuels of three basic categories. *Primary energy* is derived from sunlight, wind, flowing water, and atoms. *Biomass energy* is based upon plant growth and photosynthesis, and is renewable. We eat biomass, feed it to animals, and sometimes burn it, though in much smaller total proportions than in the previous millennia. *Fossil-fuel energy* comes from epochs-old biomass cooked and compressed into energy-dense matter. Every day, global industry burns fossil fuels that were millions of years in the making. Fossil-fuel sources are depleting, and they are famously dirty, especially coal.

Like predators in an ecological community, fossil fuels are the scarcest fuel because they rely on so much raw biomass and deep time to develop. They are highest on the fuel chain: secondary consumers of primary solar energy. Plants are first consumers of that energy, annually fixing solar energy into carbohydrates using photosynthesis, energetic plant mass that we, in turn, burn or eat. Each unit of fossil fuel requires a large platform of biomass, which is in turn built upon floods of primary energy from the sun striking our planet.

It seems natural that a literary study of fuel would follow the historical arc that has directed human history. Lewis Mumford's influential periodization of energy discerns three major eras that softly echo this book's divisions. He begins with the eotechnic period, c. 1000–1750, where wind and water were the prime movers and wood was the favored building material; next is the paleotechnic phase, c. 1750–1890, which used coal as the fuel, steam engines as the prime movers, and iron replaced wood for infrastructure; last is the neotechnic period (after 1890), when petroleum entered the scene, joining coal-fired electricity as the prime mover of advanced industrial culture (Nye 251).

At first there was water and wind, then humans yoked populations of plants and animals, and at last came the technology to exploit the fuels cached deep inside the earth, which was an event so significant that it caused its own series of distinct eras. However, the story is more complicated than this neat historical arc. Biomass has always been on the fuel scene. Writing is only a few thousand years old, coming with the advent of agriculture, but prehistorical cave paintings by hunter-gatherers depict prey animals: biomass-made-meat. Pastoral literatures of the Greek classical age that depict shepherds dancing on the grass and Bacchus crushing grapes also reveal the importance of biomass for animal fodder to premoderns. Primary energy, derived directly from sun and wind, actually came comparatively late on the fuel scene. It would be a long time after the advent of agriculture (at about 12,000 BC) before the technologies of sailing ships, water wheels, and wind mills would become the fuel sources of civilization (about 5000 BC, 300 BC, and AD 100, respectively).

As MacDuffie and other scholars such as Frederick Buell and Imre Szeman have noted, energy history is filled with epochal shifts in which a new source (coal, petroleum, nuclear) is initially hailed as a savior from a limited and problematic older source (wood, coal, and petroleum, respectively). The new source is figured in popular discourse as functionally infinite and, therefore, millennial: it promises to liberate humans from the physical yokes of material limits and hard work with virtually no side effects. Before long, usually within a few decades, problems with the new source become manifest; these problems may be technical, supply-based, and/or environmental in nature. Frederick Buell has called this a cycle of "exuberance and catastrophe," and convincingly studied petroleum culture under this rubric, noting astutely that the bipolar emotions often exist simultaneously, as they do in our present culture. In the stage of catastrophe, futurist rhetoric moves on to another source, real or imagined (remember George W. Bush touting hydrogen fuel cell cars?), and the public discourse indulges in another round of idealization worthy of science fiction. Indeed, science fiction, a genre that uses imagination to generate new ideas later to be tried out technically, is often the nascent rhetoric of exuberance. Crucially, Imre Szeman notes that the "fiction of surplus" enables the tailing, even more dubious notion that easy access to abundant sources of fuel plays "a secondary role in history by comparison with human intellect and the adventure of progress" (324). Literary fictions are collusive in both obscuring fuel as physical matter, and enabling the pure fantasy of the energy cornucopia (324).

On the contrary, a running thesis in this book is that a culture's material fuel source opens a landscape of possibility for the kinds of ideas, ambitions, and progress in which that culture can engage. This is the nature-culture of energy: resource supply and consumption. Scholars in the energy humanities call this a new form of ontology, and have committed the most focus to oil ontology (see Hitchcock, who cites the 2009 MLA special panel on oil ontology, and Patricia Yaeger's 2011 *PMLA* column on energy literature [126.2]). Without proceeding fully to an abject fuel determinism that would claim that culture is utterly determined by fuel sourcing, not human choices, it is essential to consider how the supply of fuel available has an inexorable influence on culture. A broad arc of infrastructure, social relations, values, and environmental impacts characterizes different fuel cultures through time. Without coal, there is no steam engine, no mass production, no culture of compulsive mass consumerism, and much less climate change. Without oil, not only would our living landscapes look entirely different (probably organized around steam-driven railways and muscle power), but our cultural values, ambitions, fetishes, and fantasies would shift as well, flashing on the fuels that drive the culture. To the medieval charcoal burner, a Toyota Prius would have the same use value as a Hummer: the nominal value of shelter. Postapocalyptic and peak oil narratives often engage in a proleptic view of a world drained of oil that must take stock of its obsolete apparatuses: cars and their infrastructure in particular. These husks of obsolescence are the inspirational materials of "thing theory." Without oil, there are no cars or highways, but there is also no internet age and no twenty-first-century culture of viral self-absorption. Oil ontology gives each individual consumer the illusion of being superhuman. Oil as an emancipation from the severe limits of a biomass world is a form of apotheosis: it transforms our physical ontology from bio-muscular to techno-mechanical, a transformation that brings superhuman capabilities. Your car can exert forces a hundred times stronger than a traditional pair of carriage horses, and a thousand times stronger than your spindly legs. We are masters of oil's raw power, but we have signed a contract that works in reverse, too: because of our reliance, oil masters us. Our use of fossil fuels has completely transformed the world we inhabit, and dependence creates a love/hate dynamic familiar to late industrial culture: love the car, hate the traffic; love the cheese doodles, hate the obesity; love Instagram, hate work emails.

David Nye's *Consuming Power* is a classic inquiry into how shifts in fuel sources catalyze paradigmatic changes in the societies that adopt them. Nye's cultural history of energy rejects the determinism that would claim that the evolution of fuel and technology is an unyielding force moving through history, with impacted societies helplessly enthralled by its power. Instead, Nye explores the normativity of energy in particular historical moments and how "each energy system *has been used to shape* distinctive domestic patterns, work routines, urban structures, and agricultural methods, imparting particular rhythms and contours to the everyday round" [italics mine] (8). An intriguing aspect of Nye's negotiation between a society's selection of fuel and its design of the technologies that exploit it is his assertion that there is a historical progression from freedom toward inflexibility. Later stages become rigid because societies build expensive infrastructure to support a particular fuel choice. Large-scale systems begin fairly flexibly, with a range of choices in ownership, control, and technical aspects, but rapidly grow resistant to dissidence. Nye cites seemingly small choices like the arbitrary selection of rail gauge for standard tracks, and large ones like the dismantling of American public rail transportation in the middle of the twentieth century in favor of the automobile, as examples of huge infrastructure investments that are nearly irreversible: "A generation later it is difficult to undo such a decision" (3). Happenstance looks like destiny from the perspective of the future. Evolutionary theorists like Stephen Jay Gould underscore how biological evolution can seem teleological—driving toward some predetermined goal, or *telos*—but in fact evolution is radically contingent upon environment and natural disaster. Analogically, fuel paradigms that seem destined toward improvement—more energy per capita—are the result of numerous choices based on circumstance. Many of these choices are arbitrary or selfish, intended to benefit the private stakeholder rather than society at large. In this book, a nondeterministic historiography recaptures the contingency inherent in all pathways forward.

Societies make choices about which avenues of energy and technology to pursue. American car culture was no inevitable juggernaut predestined to vanquish the railway; it resulted from a series of material conditions such as the availability of cheap oil and assembly line manufacturing, as well as a series of choices such as GM's choice to purchase and dismantle existing streetcar systems in dozens of cities, including Los Angeles, Baltimore, and St. Louis, in the 1940s. Choices that benefit the profit margins of individual corporations at the expense of the greater good riddle the history of fuel culture, with an illustrative case in point emerging in 2015 with Volkswagen's fraudulent "defeat devices" giving its "clean diesel" cars false environmental sheen. With these histories in mind, we regain perspective on the kinds of choices that will, in the twenty-first century, guide our adaptation to the material conditions of our growing reliance on diffuse primary energy: solar, wind, water, and nuclear. This nexus between determinism and choice will drive social and environmental changes through the twenty-first century. Today we have more options in energy resources than humans have ever known, but not all options are equal. Most often, our choices continue to favor retrenchment in fossil fuels and continual profits to huge corporations (seven of the top ten largest corporations are in the oil and gas business) that keep the ideology of endless economic expansion in fine clothes. Our freedom to choose alternative energy sources cinches down with every choice in favor of industrial fossil–fueled business as usual.

Because literature, as a human practice, is inherently anthropocentric, it might seem to be a weak or misleading source material for an ecocentric study of energy through history. However, the rich and under-explored vein of fuel materiality helps the sensitive reader get closer to an understanding of how energy ontology—the nature of surviving, working, and consuming—changes according to particular eras of history and under specific fuel regimes. By looking at preindustrial and "innocent" industrial literature from our position in the Anthropocene, we can reread its action and emotion with a circumspect view of the implications and effects of living with particular kinds of fuel.

Using ontology in ecocriticism may raise the concern that high-theory metaphysical ontology, which explores the basic nature of existence and reality, replaces applied humanist work in the ethics, politics, and cultural relativities of eco-literary thought. These grounded approaches that focus on particular oppressed peoples and landscapes has characterized the last decade, at least, of ecocritical scholarship. However, far from eliding the nexus among culture, nature, and economy that guides the vastly diverse global experiences of a lived environment, this foray into energy ontology is aimed at more specifically capturing how the material and cultural contexts of particular historically embedded energy sources have intrinsic effects based on class, race, geography, gender, and justice. Energy ontology is less about the abstract musings of the empowered sitting atop a pile of fossil-fueled privileges and more about conducting revelatory contextual scholarship that places the Anthropocene consumer in an enlightened history of evolving energy regimes. An essential aspect of that history hinges on the lives of laborers, producers, and bystanders, who often bear the brunt of environmental injustices based on socioeconomic disparity.

IV Energy ontology

In using the word ontology, I do not mean to invoke the metaphysical definition of the study of the kinds of entities that exist in the universe, which held sway in ancient Greek philosophy. Instead, my meaning partakes of the more recent tradition, beginning with Husserl and Heidegger, of calling on fundamental ontology to define the social nature of human existence: its qualities and quantities, its community, its physicality, its environment, and its ethics. Within his early ontology, Heidegger explores *Dasein*, the mode of being of an intrinsically social human who functions with a nascent sense of the structures of being-with (*Mitsein*) that make particular ways of living possible. In John Haugeland's words, *Dasein* is "a way of life shared by the members of some community" (423). Our shared way of life today is starkly different from the shared way of life in the eighteenth century, mostly due to the primary cause of fuel sourcing. Secondary and tertiary effects reflect back on the primary one: higher population levels (there are ten people on the planet today for every one in 1750) are supported by industrial agriculture, mechanical construction, and modern medicine; occupations have shifted from biomass to fossil-fuel based work, from farmer and hostler to truck driver and computer engineer; entertainment that used to be self-generated, such as community dances and folk music is now most often the product of industries that entertain passive consumers. When we binge on Netflix, we are consuming the fossil fuels poured into every stage of the process: development, writing, production, distribution, marketing, and of course, the petroleum-wrought device (computer or television) and its power. With renewable energies like solar and wind coming online, in time we will see the fossil-fueled load redistributed to these renewables, but the paradigm of a fossil-fueled ontology remains largely unchallenged. Fossil fuels cultivate a paradox of voracious passivity.

Since energy ontology is intimately linked to fuel source, energy scholars in the humanities regularly use the concept of ontology to generate insights related to the industrial human condition. For example, Stephanie LeMenager describes artistic reactions to the BP oil spill of 2010 by suggesting that these protests "locate the human, as an ontological category, within industrial-era infrastructures that may be not only in the throes of failure but also predetermined to destroy basic conditions of (human) living, such as water systems" (105–6). Fuel does not lie in the background or in abstraction; fuel sourcing actually designates the polarities of local and global, slow and fast, boring and exciting, sustainable and thanatopic. The more energy dense the fuel source (read: petroleum), the closer energy ontology lies to the second of these parallel terms: global, fast, exciting, and driving toward death.

It is also curiously paradoxical: perhaps because of its ubiquity, petroleum tends to vanish from cultural consciousness. As Imre Szeman notes, "The fact that literature in the era of oil has little to do with oil doesn't negate the value of energy periodization. On the contrary, one of the most valuable functions of this schematic in our present moment—or, indeed, perhaps even in the whole history of literature—is to bring to light a foundational gap to which we have hitherto given little thought. This gap is the apparent epistemic inability or unwillingness to name our energy ontologies, one consequence of which is the yawning space between belief and action, knowledge and agency" (324). In this context, energy ontology is related to Bourdieu's notion of *habitus*, where the habits of individuals and groups become imbedded in society's structure so intrinsically that they designate new realities. From the structuralist view, individual works of literature are marked by the larger eco-social structures that support them, and *habitus* shows how the structures imposed by fuel regimes lead to self-defeating human behaviors such as compulsory consumerism and sedentary living.

Literature has long downplayed fuel's material cultures while propping up the social construct of human energy, epiphany, and triumph. However, there is a rich and under-explored vein of fuel materiality that helps the sensitive reader get closer to an understanding of how energy ontology, that is, the nature of surviving, working, and consuming, changes according to particular eras of history and under specific fuel regimes. Energy density drives us toward dissonant emotions, culminating with Stephanie LeMenager's insight that "petromelancholia" involves not only an emotional life of loss, failure, guilt, and fear in our repertoire of a shared oil culture, but also that "decoupling human corporeal memory from the infrastructures that have sustained it may be the primary challenge for ecological narrative in the service of the human species' survival beyond the twenty-first century" (*Living Oil* 104). To survive, we will have to relearn how to work with our bodies to fulfill daily needs. Throughout, I pay close attention to the physical nature of specific fuel environments, especially biomass versus fossil-fuel worlds: labor as muscular and practical rather than cognitive and abstract, consumption as material (warmth and food) rather than virtual (information and entertainment), life as textured by seasonal rituals of survival (sowing and reaping) rather than the blended gray-scape of fossil-fueled constants (shopping at the grocery store).

LeMenager's gently apocalyptic idea is that our bodies will have to adjust to a pre-fossil-fueled ontology, where our muscles exit the atrophied space of the car and airplane and walk back into the evolutionary landscape of physical work and practical ritual. The biomass section of this study dwells within the fuel ontology of the working body and the seasonal textures of existence, casting a critical eye on the idealisms of the pastoral, which are themselves often nostalgic constructs of the early industrial era. Still, a biomass ontology affords a more direct translation between work performed, physical satisfaction, and subsistence in well-known natural environment, a condition that ecocritics figure as the trope of dwelling. No wonder that apocalypse narratives like James Howard Kunstler's *World Made by Hand* often cradle reactionary ethics against fossil-fuel culture, examining what we might gain by going retrograde, back to a biomass world. Racist and sexist notions accompany his vision, and call into question how much of our industrial-era gains in human equality are built on fossil-fuel ontology.

A related psychic condition to melancholia comes with our realization of what it means to live in the Anthropocene, in which human impacts on planetary systems are so profound as to be considered a distinct geological era. Timothy Morton's theory of "hyperobjects" elaborates a postmodern ecology of climate change in which human history and terrestrial geology coincide, with the horrifying, sublime effect of ultimate inscrutability (*Hyperobjects* 12). Timothy Clark's study of the Anthropocene also suggests that within this new ontological order, our ability to interpret natural patterns using literature is highly attenuated, since imponderable phenomena like climate change "resist representation at the kinds of scale on which most thinking, culture, art and politics operate" (x). Anthropocene scholars have established how the eco-cultural condition in which we conduct our lives is marked by the wholesale conversion of nature into materials that energize human economies. This extractionist mindset is responsible not only for widespread habitat destruction and climate change, but also marks a poignant psychological moment in cultural history when denial and self-destruction become normal behavior within economies (Clark xi). Perhaps Dr. Seuss captured it best with his Onceler, who steadily consumes himself out of a profitable business by denying the resource limits decried by the Lorax. He chops down all the Truffula Trees and ends up alone in a wasteland. That a children's story written in the 1970s could be as wrenchingly sad as it is popular is a testament to the extent to which Anthropocene is already embedded in our collective conscience.

Steeped in a culture of crisis and indulgence enabled by fossil fuels, in a geological era inscribed using fossil-fuel fonts, it is worth questioning whether there can be any clear readings of fuel literature written by people who lived in pure ignorance of the larger effects of their combustion. Alan MacDuffie, Jesse Oak Taylor, and other nineteenth-century scholars have discovered some early awareness of the global effects of industrial emissions, particularly among prescient cultural critics like John Ruskin in his jeremiad "The Storm-Cloud of the Nineteenth Century," but those portentous moments are the exception (MacDuffie 169). To read authors whose characters are warmed by coal, like Jane Austen, Charles Lamb, Charles Dickens, and Charlotte Bronte, in this latter-day context of cause and effect, we must acknowledge that the readings in this book present a kind of self-conscious Anthropocene rereading of literature. They knew not what they did, perhaps, but we cannot pretend we know not what we do. One key quest of this book is to find the moments in which the consumption of fuel became self-reflexive; that is, when art began to represent what we might call an Anthropocene awareness that human activity had the power to change geo-global history. The British nineteenth century is the era in which these observations shifted from innocence to awareness.

Even though the skies above London have mostly cleared since their time, dispersing the dour industrial ambience of *Bleak House*, today we cannot help but read onto those early works our own hypertextual version of ontological conditions caused by fossil fuels and the Anthropocene: guilt, melancholy, fear, helplessness, and, despite it all, exhilaration. It has often been noted that environmentalism suffers from its shrill, lamenting voice in which elegy and jeremiad match their decibels. Comedy is rarely heard outside of bitter satires on the modern human condition, and Polyanna narratives are anathema, because they appear to support a cornucopian capitalist agenda (Branch 379). By looking into literature penned before the Anthropocene was a known concept, we may recover some of the levity of innocence and optimism that colored preindustrial literature, and the blushes of infinite possibility unleashed with coal and petroleum.

Slinging nightmares, environmental writing works against its own interest in rousing popular sympathy, even with the spectacle of apocalypse; presenting dreams, our work supports the larger investment in imagining how a better, more sustainable future might be crafted out of the best iterations of pastoral, technological, and collective. It falls somewhere between dumb and dangerous to oversell cultural representation as a force of change in human behavior (Clark 21). However, the energy humanities can help elucidate ecological problems for the public, especially by identifying how tropes and narratives inhere in our creation of nature and human culture, even if our work contributes little to the direct application of science and policy that helps avert, assuage, and minimize negative impacts of ecological crises.

The exhilaration of living in the Anthropocene is perhaps the most complex among a clutch of complicated feelings. Our culture is exhilarated by the godlike power at our disposal, what William Catton Jr. described as *Homo colossus* (just imagine, as Upton Sinclair did, a train of three hundred horses galloping in advance of a carriage!), but it is also exhilarated by the pure uncertainty, scientific and literary, of what beast of climate we may have unleashed, and how these changes will play out within our lifetimes and beyond. Where the biomass literature of haystacks, charcoal burners, and wood piles belongs in the pastoral tradition, fossil-fuel literature partakes of the apocalypse. The search for intelligibility may be inherent in the very practice of rereading fuel literature in the Anthropocene. If our contemporary culture is "on a darkling plain, / Swept with confused alarms of struggle and flight," as Matthew Arnold wrote in a Victorian plea for sweetness and light, looking back along the path of our predecessors may provide us coordinates for our own situation. The singularity of literature, to use Derek Attridge's phrase, is its intrinsic ability to present a unique and innovative Other that enriches a culture's ethical terrain, but that the culture itself cannot immediately assimilate. The material fuel that colors the experience of literary generations before us is bound to shed light upon our condition and, from the Anthropocene perspective, we may be able to attain more insight on the significance of fuel regimes of the past (Attridge 2).

The material fuel at work within these texts is hardly ever its central focus, which overtly circles more celebrated literary topics like triumph and tragedy, love and indifference, selfishness and altruism. But the material fuel in the background actually performs the fundamental work of drawing the landscape of possibility for the characters acting within its horizons. Tim Clark has identified scale effects in climate change, where "numerous human actions, insignificant in themselves … come together to form a new, imponderable physical event, altering the basic ecological cycles of the planet (72). Background—the material fuel—becomes foreground as its impacts accumulate. Rereading biomass, fossil, and renewable fuels yields a review of historical literature—a preface and rationale for engaging in the primary research of living this twenty-first-century life with the fullest possible awareness of our own horizons.

Incredibly, our industrial culture has mostly forgotten what it was like to exist in a biomass world: sustainably, locally (even parochially), and slowly—no faster than a horse's gallop. To consider the comparative literatures of energy as they relate to transportation, we will begin with wind, then consider muscle-driven biomass giving way to coal locomotion, and conclude with the highest octane fuel, petroleum. My interest is in how the fuel depicted in literature illuminates historical moments in which the interfaces between self, society, and nature are configured by specific energy regimes. By using literature as a source text, we may arrive at an emotionally and philosophically more robust synthesis of energy history than the social and natural sciences, relying upon objective accounts and statistics, are able to provide.

V The arc of *Fuel*

Because its historical scope is broad, *Fuel* provides an abridged review of fuel literature with a running commentary on current issues of landscape destruction, pollution, biological limits, environmental justice, and climate change, all of which relate to our position in the Anthropocene. It would be absurd to try to capture every literary puff of smoke within these pages, so I have selected works that best transmit the insights on ontology that we seek in our time of overstimulated befuddlement. The primary intended audience for *Fuel* is college students and curious adults who do not necessarily work in academia. I have intentionally used a little specialized language as possible so as to attain a wide-ranging historical scope that engages general readers.

The book proceeds intuitively, with major sections on biomass, fossil, and new renewable fuels, each partitioned into specific contemporary interests like wood fires and hospitality, coal and rapid locomotion, the split atom and the fallout fate. There is a diversity of genre, from poetry to fiction to essays, despite a generally traditional view of what constitutes "literature" (I have not included readings from the social or natural sciences, for example).

Fuel is a book of literary exploration, so it is slender on statistics and quantitative analysis. Historical economists have produced many worthwhile studies of fuel and energy through time. Particularly notable is the work of Vaclav Smil, specifically his book *Energy in World History*, a primer on the evolution of culture and technology based on shifts in fuel sourcing. For example, Smil's data and analysis show how the value of working animals on preindustrial farms depended intimately on the amount of biomass that could be spared to feed them. Large numbers of horses and oxen could subsist only in areas with great expanses of arable land, like the United States and parts of Europe; areas poorer in land relied more on human labor (73–80). Smil also usefully considers the roles of charcoal, windmills, and waterwheels in early industry, and how these primary energy and biomass architectures were torn apart to make room for coal in the nineteenth century and petroleum in the twentieth (164–5). I am also indebted to work by cultural historians like Richard Wrangham, E. A. Wrigley, Frederick Buell, and Barbara Freese, among others, and you will see traces of their narratives in the present work. These scholars have been especially astute in their analyses of the Industrial Revolution, with coal's lead role in affecting myriad changes in environment, self, and society.

In addition to social and ecological histories of fuel, energy scholarship has recently turned to literary sources, which add a complement of fictional history, emotion, imagination, and ontological questioning to the social sciences. The recent work of Bob Johnson and Stephanie LeMenager has opened the world of fossil fuels to literary cultural analysis, with especially powerful insights on the physical and mental traumas of carbon-era living, the general disembodiment of labor ironically juxtaposed to the subaltern bodies toiling to produce coal and oil, and the narrative trajectories of expansion and declension that guide our cultural perception of energy. Johnson's *Carbon Nation* and LeMenager's *Living Oil* are excellent investigations of fossil-fuel ontology, as is the recent writing of Patricia Yaeger, Ken Hiltner, Michael Ziser, and Imre Szeman.

This book's scope is broader than oil and coal, intending to compare the ontology of the "in" group, fossil fuels, with lives run upon the solar economies of biomass and primary energy. Though the three categories of fuel fall somewhat into a historical arc, they all coexist as well, and their exchanges and overlaps are the nexus of literary insight. This book departs from traditional sociohistorical inquiry to provide some ideas about how poetry and novels—fictional perspectives that encompass, amplify, and elaborate upon reality—are unique vessels for considering fuel. The literary container is potentially experimental, imaginative, speculative, political, ecological, and activist-oriented. Literature is a gumbo of history, philosophy, art, science, and politics, a stew of interdisciplinary knowledge that creates new intellectual tastes to savor. Because literature is so complex, it has the capacity to bring a background issue such as fuel into the foreground. As an inherently reflective medium that supports metanarrative, a counter-reading of literature as a source text for understanding fuel ontology can be particularly timely and rewarding. What I wish to do here is consider recipes using the ingredients of aesthetics, emotion, identity, and, more fundamentally, human ontology, as they relate to fuel. In our historical myopia, we often lose sight of life before the reign of fossil fuels, which began only two centuries ago. What is hidden or taken for granted, I wish to expose. However, because fuel became such a dominant force in culture during the industrial age, the literature of this book is highly concentrated on writing since 1800.

Our language is inflected with metaphors of energy. Older phrases like "three sheets to the wind," "horsing around," "feeling your oats," and "nose to the grindstone"—elements of a preindustrial world—are close to falling into the realm of dead metaphors. They have given way to the industrial metaphors of productivity based on the steam engine, where we "let off steam" so we don't "blow a gasket" (Nye 90). Our metaphors have changed from designating a world of animals, human bodies included, to designating a world of machines that have appropriated the body's work. The language implies that your cranium is now more like a gasket than the head bone of an ape.

It has been surprisingly easy to slough off our evolutionary and physical proximity to other animals and embrace machines as our kinfolk—and even engineer them to have autonomous agency and intelligence. Our concern for machines approaches sentimentality: likening the new locomotive to a "snorting little animal," one of the first rail passengers "felt rather inclined to pat" the contraption, as she would her trusty mare. We "baby" our cars, "debug" our computers, and often spend more money and emotion on their maintenance than on creatures that are alive, and suffering. Extending from the life of the metaphor and its power to construct meaning and identity, today we find ourselves literally, materially, made of oil products. This metamorphosis from animal to enhanced petro-animal happens most directly in food, which today usually consumes more fossil-fuel calories than it provides in animal/vegetable ones. It also happens ontologically as we extend our animal selves using energy-intensive prosthetics—things like cars and computers—that help us move and think better, or at least faster. Fossil fuels flummox us into an alluring and dubious "faster is better" ideology.

Sources of fuel, biomass like wood and peat, fossil fuels like coal and petroleum, and renewables like hydropower and geothermal energy, might seem mundane. Is not the interest of literature found in its emotive actions, not its dumb nouns? What literature teaches us about the human condition usually comes from the psyche of complex characters, not from the matter that drives their physical lives. For most humanist scholars, the paramount interest is the interplay between inner psychology and outward action, agency and fate, history and chance. Often literature elides its literal fuels, instead preferring the metaphorical fires of love and anger, the changes augured by the west wind, the depths of coal-black eyes, and the stoutness of oak hearts. Though we generally attend to the tenor, in this book I turn our eye on the vehicle of fuel metaphors. *Fuel* contends that the very nature of place, action, emotion, and fortune is in a large part designated by the material energy powering the scene.

Because of the prevalence of fuel in literature, I have had to select the works that best evoke the human condition within the rubric of a particular fuel, and leave out many others. The literature sampled here is generally Anglophone, with the insight of the fuel source driving the selections rather than any desire to focus on particular nations or genres. There is wide inclusion across class, race, and ethnicity, and readers will recognize well-known canonical works alongside the more obscure but no less illuminating selections. Following a historical arc that is not nearly as neat as the rainbow, we will begin with the oldest of human-used fuels, biomass, by considering the fodder of grass and hay that was critical to agricultural humans, and the pastoral rhythms that energize literature, starting with Virgil's shepherds and continuing all the way to a postapocalyptic pastoral world envisioned by German writer Marlen Haushofer. It may surprise the modern consciousness to begin a book on fuel with fodder. The bias that conflates fuel with fossil fuel is one of the weak points in ecocritical engagement: energy is often treated only in its most recent form, which catalyzed the Industrial Revolution. But fodder is the original fuel for the agricultural human, and we have labored to produce it for working animals ever since we had fields to plow and cows to milk.

Next, fire-giving wood and charcoal fueled innovations of cooking, indoor heating, and early industry, and the importance of the forest-dwelling charcoal burner comes center stage through its rather obscure literature, from the children's writer A. A. Milne to the creole hollers of Lafcadio Hearn. Finally, we turn to more unusual biomass sources like whale oil and tallow, animal products essential to pre-fossil-fuel illumination, and peat, a "slow renewable" heat source used intensively in regions with bogs, particularly Ireland, that claims its own leitmotif in the poetry of Seamus Heaney. With each of these biomass sources, themes that defined human existence since the agricultural revolution more than ten thousand years ago will be important to our understanding of biomass ontology: hard physical work, attunement to season and weather, synchrony with working animals and other laborers, and the perils of shortfall, particularly failed harvests and deforestation.

Next we will leap into the industrial space-time of high-energy fossil fuels, which have received the most attention in the energy humanities for good reason. Coal shouldered the Industrial Revolution beginning in the eighteenth century, but had been used more sparingly for heat centuries earlier. Its singular power to drive the steam engine revolutionized the means of production and distribution, requiring higher agricultural yields of raw goods such as cotton, as well as luxury items formerly consumed only by the very wealthy: tea, coffee, sugar, and chocolate. All of these products were grown mainly with slave labor in the Americas. In the imperial center, Britain, coal transfigured landscapes, both urban and rural, and created the possibility of suburbs along railway lines, reshaping human settlement patterns and the kinds of work performed by all classes, from rural peasants, who flocked to cities to become factory workers, to the wealthy classes, in which ancient, landed wealth made way for industrial barons quickly enriched by the new consumerism. Coal also supported a society of middle-class workers who could aspire to such aristocratic luxuries as large homes with new goods to fill them, higher education, and leisure travel. The environmental, epidemiological, and emotional impacts of widespread coal use are intricately recorded in nineteenth-century literature, perhaps most indelibly in the works of Charles Dickens, but also in more subtle ways in the writing of the Romantics, the Brontë sisters, and Anna Sewell's bestselling *Black Beauty*.

These texts are the first wave of reckoning with the fossil-fuel ontology that characterizes our industrial experience. They sit at the pivot point in history when fuel use was becoming self-reflexive—the nascent awareness of living in an Anthropocene.

The next fossil fuel, petroleum, is a less striking paradigm shift than biomass to coal, but oil held the power to modify rail-based systems into the private car—that classic symbol of freedom that doubles as its own fuel ontology. Particularly in America in the twentieth century, individual rapid movement redefined time and space on the map and the mind. Jack Kerouac is one among many high-octane American writers who explored the thrill and peril of a high-speed, transient existence characteristic of an age of fossil-fueled fury and ennui. Oil products also served as stock material for whole new categories of consumer goods made from plastics. The age of the internet was made possible, materially and energetically, by oil. The fossil-fuels section also contains short excursions into less pervasive materials such as coal gas and natural gas, the latter of which is enormously important in this era of hydrofracking and transitioning away from coal to generate heat and electricity. Natural gas has left only a faint trace on literature compared to its kindred fuels. In a short interlude that partakes of all categories of fuel, we will consider human fuel—food—in its pastoral and industrial versions, particularly in the poetry and philosophy of homesteader Wendell Berry and Billeh Nickerson, a McDonald's employee.

The last major section of the book turns to primary energy derived from the sun, wind, water, and atom—sources of energy with both ancient and contemporary lives. Ambient energy from natural flows like wind and water held great importance for preindustrial societies, whose mills refined wheat berries into flour and performed other grinding tasks like the preparation of gunpowder. These energy sources varied dramatically according to natural conditions, and early feats of engineering regulated water flows using millraces and containment ponds. Still, drought and flood and, for windmills, calm days, often suspended production. George Eliot's Tulliver family is helpless in the face of the elements—an ontology in common with biomass societies that relied on the harvest, but one quite foreign in our empowered age of fossil fuels, which can be deployed regardless of natural conditions.

The modern era of the giant dam began with the public works projects in the mid-twentieth century United States, and this great elevation of dam-building scale and ambition showed a fossil-fuel era's approach toward regulating natural water flows to control flood, opening rivers to large shipping vessels, and generating electricity. The effects on riparian fisheries and tribal fishing rights, as well as the landscapes flooded in the wake of the dam (most famously Glen Canyon), continue to be controversial sacrifices in the name of progress. Writers like Edward Abbey are synonymous with dam protest, and his work in particular shows how fiction can inspire real-world activism. Though large dams continue their furious pace of construction in developing regions, the era of dam decommissioning is well underway in most highly industrialized countries. Because so much literature might be included in a study with such a broad scope, many of the readings are short and pithy. However, more in-depth studies of a few remarkable works appear in each section.

It is important to remember that we hold the benefit of hindsight when reading the literatures of previous centuries. Fully aware of the global impacts of the Anthropocene, and particularly the changes of last two industrial centuries, it could appear a form of critical bad faith to foist judgments upon the behaviors of past generations. As much as possible, I have placed perception over judgment, and left the ethical orientations of the writers themselves to stand as a historical record of how literary figures marked their own experience. My interest is in how the fuel depicted in literature illuminates historical moments in which the interfaces between self, society, and nature are configured by specific energy regimes. By using literature as a source text, we may arrive at an emotionally and philosophically more robust synthesis of energy history than the social and natural sciences, relying upon objective accounts and statistics, are able to provide. Rereading these texts with an emphasis on fuel helps us mark today's cultural and ecological coordinates. Tomorrow's literature will continue to record the evolution of fuel cultures while developing characters whose ontology reveals the immediate and downstream effects of consuming energy. The literary imagination pays forward its insights.

Part One

Biomass

2

Grass: Muscle Power

For millennia before it was powered by steam, internal combustion, and nuclear fission, the human world ran on muscles. Initially we ran on our own muscles, powered by meat from hunting and tubers, nuts, and fruits from gathering. One of the bracing advances of the agricultural revolution was the extension of our own modest frames to the much stronger bodies of ungulates—animals that walk on hooves—like horses, oxen, and donkeys. These tip-of-the-toe animals run on rough biomass. Graminivores, especially ruminants, are marvelous alchemists that turn things inedible to us, mostly grasses, into power, speed, meat, and milk. (Horses are not ruminants, so they get less energy from grasses than cattle and require higher-calorie cereals in their diet.) Our association with these animate prime movers is difficult to overstate. They shouldered a series of revolutions in global human civilization: fixed settlement, plow-based agriculture, which led to a shift in human diet toward carbohydrates from cereals, huge population increases, the advent of the city-state, class-based hierarchy based on occupation, and beer. How now proud cow?

In a world presently so dominated by fossil fuels, it is easy to neglect all the sources of power that preceded it. Biomass and elemental energy crafted the civic human world from the raw elements of wilderness long before there were steam engines. Without the wind, no sailing ships that enabled preindustrial global exploration, imperialism, and trade. Without rivers, no mills grinding flour and gunpowder, and no lumber. Without strong animals fed on biomass, humans' ability to travel across land and make arable fields from forests by pulling out stumps, for example, would have been greatly reduced (Nye 19). Though fossil-fueled technology has transformed nature and society to a greater degree than any other force, it is myopic to neglect the impact of preindustrial sources of energy. Muscle power driven by biomass first enabled settlers and colonists to convert nature into natural resources, to translate ecology into economy, and to lay the ideological groundwork for an economic structure, capitalism, that views organic nature as a store of commodities.

By contracting out our muscle work to other animals, we humans were freed from certain burdens, such as transporting ourselves and our goods, but we were yoked to numerous others. Animals needed tending, feeding, shelter, breeding, and healing—thus the invention of the occupations of ostler, driver, groom, farrier, and veterinarian. With animal domestication came shared diseases. The beasts deposited nitrogen-rich dung, a boon and a pox depending on context: as agricultural fertilizer, a boon; as disease dish and source of pollution, a pox.

In spatial and ecological terms, rural animals trod more lightly than urban ones, who needed their food harvested and imported to the city, and liberally dropped dung in their cobblestone path. But the hard work of farm animals came at the cost of providing them with high-quality biomass fuel. Hay and straw provide stomach-filling fodder but not much energy for large animals. In order to work hard, horses needed wheat bran, oats, and legumes. The twenty-five million horses living in America in the first decades of the twentieth century required one quarter of all arable land to grow their grain feed (Smil 73). Because of this burden, only countries rich in agricultural land could afford to support a large population of working animals. A country's agricultural development depended upon the number of graminivores its land could support.

There were always trade-offs between keeping animals for work (and fertilizer, milk, and meat) and maintaining a more energetically modest human-driven farm. Thinking of horses and oxen in terms of energy exchange helps us understand what Bob Johnson as called the "ecological cul-de-sac that defined the premodern economy," and how fossil fuels meant liberation from the limits of the land (20). Again, as with automobiles for transportation, the arrival of coal and petroleum to power farm equipment represented a major relief from the demands of cultivating biomass to power muscles. The fact that these incredibly concentrated fossil sources were nonrenewable and produced a new kind of air pollution mattered little to the first industrial generations.

This chapter is devoted to the literature of biomass-fueled beasts and the world that was plugged into their muscles. Though we tend to think of journeys on horseback, teams of yoked oxen, and asses pulling wagons as long gone in the modern world, these animals continue their steady work in undeveloped locales across six continents. Livestock continue to be the stock holdings of family wealth, especially in regions where habitation would be impossible without the cow's milk and meat, as in the Kalahari Desert in southern Africa, or the Andean highlands of Peru, from which a large portion of alpaca wool originates. Ungulates well-adapted to earth's diverse grass and scrublands open niches for human existence in places with severe temperatures, low rainfall, extreme elevation, or little arable land. The farmer leading his laden llama down to the regional market is benefiting from the energy of the sun: first stored in grasses, then converted to muscle and released, stride by stride, by the steadily working animal. The weaver turns solar-energy-made-matter into thick rugs and soft sweaters on her loom.

Living on biomass implies certain relationships with time and space now largely absent in fossil-fuel-driven industrialized society. Unlike machines, animals must be given reasonable working times and loads, and their range and speed are likewise set within physiological bounds. There are times in history when humans pushed animals to the brink of their endurance, such as the mail-coach days of mid-nineteenth-century England, when animals came into direct competition with the new technology of the locomotive. But these extreme and often cruel regimens are not the ones followed by a prudent steward, whose animal's well-being may represent the family's ability to survive. Living on biomass implies slower time, tighter range, deeper dwelling. It rewards theriophily—love of animals and appreciation of their essence—because skill, rapport, and affectionate care are foundations of a working synergy.

As Wendell Berry remarks in *The Unsettling of America*, agricultural skills suffered "radical diminishment" when farms flocked to machine-based (fossil-driven) labor, because with animals, "two minds and two wills are involved ... success depends upon the animal's willingness and upon its health; certain moral imperatives and restraints are therefore pragmatically essential" (92–3). On the other hand, the machine is lifeless and without will, so the moral limits are drawn exclusively, and precariously, within the human sphere. Traditional biomass energy tends toward the preservation of nuclear families and small communities in steady subsistence; fossil energy tends toward the agglomeration and dispersal of massive systems of industry, transport, urbanism, and consumption. The reason for this basic distinction is biochemical: biomass accumulates seasonally in sun-knit grasses with modest annual totals; fossil fuels pour from the ground in millions of years of sunlight per haul. Dried grasses contain about one-eighth the energy per weight that petroleum does, and though engines waste about 70 percent of fossil energy, they are still more efficient than animal bodies at converting energy into work (Smil 12). The engine does not have a resting metabolism.

John Ruskin was one of many philosophers alarmed at the onslaught of steam-driven rapid transit in the nineteenth century. Seeing the shift from muscle-based work to a more intensive coal-fueled world, Ruskin warned of the replacement of biomass ontology with fossil-fuel ontology, in which "a fool wants to kill space and kill time: a wise man, first to gain them, then to animate them" (*Works* Vol. 3 320). The animation of space-time is tied to the basic exchanges between sun and earth, vegetal photo synthesizers and herbivores. It is impossible for a skilled animal keeper to be entirely anthropocentric. Berry joins the Ruskin camp with his observation that machines replaced animals on farms mostly because of their greater speed: the ratio of amount of work completed over time. The speed creed has no morality beyond an economic one. But what is sacrificed, these pastoral philosophers argue, is restraint, skill, responsibility, and the balance between use and continuity—that is, the ability to provide for the future in one's present actions (*Unsettling* 93). An animal laborer consumes the fodder of the present and deposits the dung of a fertile tomorrow. The tractor feeds on money, purchased fuel, and so the industrial farmer must concentrate on production and sale before provision and responsible use.

It is easy to be carried away by nostalgia for the cozy images of a preindustrial lifestyle characterized by silence and slow time, now obliterated by coal and petroleum. The neurasthenia caused by modern living seems to cultivate pastoral fantasies of a better time now past. Traditionally the pastoral is linked, by way of life, to senescence and death, but even death is a gentle ideal in biomass poems like Keats's "To Autumn." Warm bronzed hay may be appealing in a painting or poem, but what about the penury of a bad harvest or an extreme winter, when the annual store of biomass energy falls short? In effect, fossil fuels and machine-driven labor represented liberation from arduous, repetitive work within the severe confines of an energy economy driven only by the sun. Wary of the idealisms of pastoral literature, I have selected a range of pieces, idylls, rants, and elegies that more accurately cover the range of conditions in grass-powered lives. Since this biomass section serves as a counterpoint to the fossil-fuel hegemony, its preindustrial ontology should shed light on the industrial structures that have become invisible when dwelling within the world of coal, petroleum, and natural gas, as we do.

Biomass occupies a position in eco-cultural history nearly as important as the industrial transition to fossil fuels that we will look into in the next section. Biomass fostered a revolution in the eighteenth and nineteenth centuries in Europe and North America. From economies of rural subsistence, intensive cultivation of natural resources converted the rural sphere into a state of agrarian capitalism in which grass and grain and the animals they nourished became commodities of crops and livestock actively traded for financial gain. These commodities represented the actual and symbolic wealth of a growing, modernized economy that further divided country from city, encouraged an exodus to urban factory jobs, and stimulated the era of capitalist consumerism that is a central feature of our lives. Though coal was necessary for these shifts to be ratcheted into the enduring cultural paradigm of industrialism, the history of biomass is a vital component of the narrative of modernization.

From fueling modest, feudal, subsistent, pastoral lives, biomass speaks to the moments when even renewable natural resources like grains were drawn up into the modern economy through the mediation of fossil-fueled machines. For all of the retrospective, pastoral perspectives on biomass, it is also a fuel with a future. Intensive research into renewable biomass energy from crop staples like corn produces energy from ethanol, which contributes up to 10 percent of the mixture we buy at the gas station. A second generation of biomass that neither competes with human food supplies nor requires intensive petroleum-based fertilizer is one of the major frontiers of renewable energy research. Perennial grasses like switchgrass, native to North America, may in the future fuel ethanol production on a large scale, while working to preserve prairie habitat so thoroughly degraded by over a century of intensive agriculture in the American Midwest.

A word on pets and pot roasts. This chapter covers the lives of the animals, humans and otherwise, whose energy comes from renewable biomass, which preceded the fossil-fuel driven modern industries that now feed our domestic quadrupeds and fatten our meat animals—particularly chickens, cows, and pigs. Using fossil fuels, we have created distinct categories of animals: wild animals, especially predators, for spectacle; domestic pets for companionship and amusement; replicate meat machines to fill our bellies. Outside of these categories lurks a time-honored, now rare, way of living with animals as equal parts companion, fellow laborer, and supper. The eighteenth-century poet Ann Yearsley held a pseudonym, Lactilla, indicating that she started her life as a milk-woman. The single line "Lactilla tends her fav'rite cow" celebrates an animal with an individual personality (she is "favorite" for a reason), an exchange value (milk), an influence strong enough to label her keeper (Lactilla—she who milks), who demands certain accommodation (she requires tending). The fav'rite cow might end life as a pot roast, but only after living a more textured existence than we now allow for wild predators, pets, and stockyard steer in the industrial state. By looking at the fuel of biomass, I hope to open out this animal ontology that has all but vanished in developed nations, and show how biomass designates a certain mode of respect, even equality, between working animals and their humans.

I The country: The pastness of the pastoral

Escorting his flock, the shepherd, hero of the literary pastoral, climbed atop the haystack of classical and early modern literature. Poems by Virgil and Theocritus, Spenser and Marlowe, are much more attuned to the aesthetics and emotions of an idealized rural society than they are to the energy sources of that world. I leave you to read the legions of scholarly tomes on the pastoral to gain an appreciation of the vogues, declines, satires, and imitations that fill literary history (see, for example, Paul Alpers's *What is Pastoral?*). For this book, we will remain trained on grass and oats, hay and horsebread. This is writing with "earth clinging to its roots," to quote Thoreau. These vegetable origins of modernity are carbohydrate chains fixed in photosynthesis by solar power, "the force," as Dylan Thomas imagined, "that through the green fuse drives the flower." Biomass passed through the guts of animate prime movers to their muscles and fueled the agricultural society that was a radical departure from the previous two million years of Paleolithic culture. The domestication of animals was the first stage in the intensification of agricultural work, followed by the non-fuel-based advances in irrigation, fertilization, and rotation of crops. The well-bred horse could perform the work of six to eight humans, and could produce more total force for large tasks like pulling out stumps and dragging large plows (Smil 39–40, 49, 86).

Figure 2.1 Winslow Homer, *Making Hay*, 1872. Image courtesy of the National Gallery of Art. Public domain.

The pastoral is a distinctive genre and setting in literature. Aristotle was a scientific-minded pragmatist who thought that nature provided cattle for mankind, "both for his service and his food" (qtd. in McInerney 1). Celtic myths are filled with cow worship, even to the point of mythic warfare over ownership. Romantic poet John Keats, who shunned the industry that transformed his London, viewed a pastoral scene of "silence and slow time" on an urn in the British Museum and asked in his *Ode on a Grecian Urn*: "To what green altar, O mysterious priest, / Lead'st thou that heifer lowing at the skies, / And all her silken flanks with garlands drest?" (ll. 2, 32–4). The scene on the urn documents the essential role cattle played in classical Greek culture, since the animal provided meat, milk, and labor. The cow was so valuable to ancient cultures that they routinely conducted rituals of sacrifice to the higher powers in an appeal to keep the grass growing, the muscles going, the milk flowing.

To Homer, the cow was significant for its symbolic and exchange values, essential to the economy of marriage as dowries, and used as a way to display wealth and denote status among herd-holding families. These exchange values were acute enough to translate into a symbolic value in sacrifice, for assuaging, and gaining favor with, the capricious Greek gods (McInerney 2). To this day, the Greeks' choice of fine cows as sacrificial animals shows their reverence for the value of the animal. They groom her hide until it resembles silk, and bedeck her in floral garlands—a ritual preserved well beyond the end of classical pastoral in works of art.

For instance, Keats continues on the urn: "When old age shall this generation waste, / Thou shalt remain, in midst of other woe / Than ours, a friend to man" (ll. 46–8). The ideals of beauty and truth conveyed by the silent urn are embodied by that lowing heifer, a symbol of the beauty of the animal and its importance to the culture and ecology of Greece. Though generations are cut down, the sacred, sacrificial cow endures as a symbol of the eco-cultural relations among working animal, human steward, and nature—or the gods that, in myth, controlled natural conditions. This basic relationship between agents and elements that constitutes agricultural survival has become more obscure since Keats's time, but the Grecian urn reminds him and us that we are not in total control of our fate, and ritual is our gesture of deference to whatever *is*. Biomass involves looking to the skies with hope and fear—a condition of vulnerability. Once industrial societies adopted fossil fuels, the machine and the mine—internal, self-referential subjects—came to replace our sky speculations.

In the Roman Empire of the first century BC, the poet Virgil, in his *Eclogues*, defines the shepherd's happiness and his ability to survive according to the quality of his biomass:

> So in old age, you happy man, your fields
> Will still be yours, and ample for your need!
> Though, with bare stones o'erspread, the pastures all
> Be choked with rushy mire, your ewes with young
> By no strange fodder will be tried, nor hurt
> Through taint contagious of a neighboring flock.
>
> (1. 58–63)

The vagaries of weather and disease that annually threatened the herder and his flock emerge as divine interventions—both as blessing and curse. The pasture is a metonym of divine favor, a blessing translated as ample, edible fields, rather than "strange fodder" that would sicken the sheep. When the gods smile, times are easy, the sun shines, the hay ripens, the flock fattens. This pastoral idyll can pivot rapidly to the opposite trope—that of apocalypse—evoked here in Virgil's *Georgics*:

> Here from distempered heavens erewhile arose
> A piteous season, with the full fierce heat
> Of autumn glowed, and cattle-kindreds all
> And all wild creatures to destruction gave,
> Tainted the pools, the fodder charged with bane.
> Nor simple was the way of death, but when
> Hot thirst through every vein impelled had drawn
> Their wretched limbs together, anon o'erflowed
> A watery flux, and all their bones piecemeal
> Sapped by corruption to itself absorbed.
>
> (3. 478–87)

Under the "distempered heavens," the stew of death filth is mixed by animals, domestic and wild, leading to a holocaust of "full fierce heat" that consumes itself. The source of death is the "fodder charged with bane" and the paradox of "hot thirst" and "watery flux"—thirst in life and oozy water in death—form a universal spongy mess, "corruption to itself absorbed." Biomass is both nutriment and bane. Through it, the gods giveth and they taketh away. Virgil's famine verse recalls the delicate balance of conditions necessary to maintain the idyll. Elysium is one drought away from a wasteland; if the rains fail the grass dies first, and its grazers flee or die.

American ranchers—modern shepherds—are well-acquainted with the dire straits of a dry season, and in droughts they sell off cattle to wetter areas. In the 2014 drought in the American west, "fields that should be four feet high with grass are a blanket of brown and stunted stalks"—a modern-day life-or-death pastoral (Nagourney and Lovett). Despite the antiquity of the classical writers, their tales of natural disaster echo in our own times. The agency of apocalypse has shifted through time, however, from supernatural to natural to human. The ancients used pastoral literature to preserve their rituals of entreaty, begging the higher powers for high-quality biomass, and to record the effects of famine and pestilence when the gods failed to favor them. In the twenty-first century, when we experience extreme drought weather, we must turn to a shifting measure of agency between natural oscillations and anthropogenic climate change. Where classical literature turned to the gods, we are left, in fossil fuel ontology, in an indeterminate space between receiving the elements and clumsily conducting them ourselves. This shifting boundary between natural and industrial climate causes is a signal element of life experience in our time, and fringe groups have looped these causations back onto theological origins, with divine retribution acting as feedback for our pollution.

By the time of Queen Elizabeth I, England had a literate class that sought diversion and education through reading, and writers acquiesced by adapting old pastoral forms of literature from the Greeks and Romans. The pastoral turned from the hard work of the georgic farm labor to the more abstract and idealized fancies of the shepherd's lascivious lingering. The Elizabethan pastoral is bursting with nature scenes, fresh seasons, frolicking maids, desire, and sex—or desire but no sex. It rarely alludes to the chemical energy that runs the scene: the grass, hay, and grain. Eschewing realism, the writers peopled their poems with an invented leisure class of peasantry that reflected and flattered the reading leisure class of courtiers and city dwellers. The idealisms of the pastoral became the source of literary fencing matches, as when Christopher Marlowe's "The Passionate Shepherd to His Love" was parried by Sir Walter Raleigh in his "The Nymph's Reply to the Shepherd." Marlowe's shepherd promises a natural jewel box of roses, posies, myrtle, ivy, coral, and amber, and Raleigh's pragmatic maid responds:

The flowers do fade, and wanton fields,
To wayward winter reckoning yields,
A honey tongue, a heart of gall,
Is fancy's spring, but sorrow's fall.

Thy gowns, thy shoes, thy beds of Roses,
Thy cap, thy kirtle, and thy posies
Soon break, soon wither, soon forgotten:
In folly ripe, in reason rotten.

(ll. 9–16)

Raleigh cleverly uses the seasonal pun of light-legged spring and stumbling fall to indicate that the lives of rustics take footing in the conditions that are good for growing grass. Shining sun, warm air, growing grass, fattening beasts, happy shepherd, and love. The autumnal opposite, the nymph points out, is both inevitable and predictable. Vernal boy has no plan to adopt when the earth pivots away from the sun. If the maid were more inclined toward his addresses, she might devise for him a rational autumn of mowing, curing hay, and a hearth fire well-fueled by split cordwood. The Christmas jingle "Let it Snow" draws out that wintery ideal. But this nymph is a contrarian, not a coauthor of a romantic alliance. Neither poem mentions the commonplace of grass, but the whole terrain of rural existence is carpeted in biomass, so, even in a poem of roses and posies, grass is bedrock.

Andrew Marvell, the near contemporary of Marlowe and Raleigh, composed a series of poems on Damon the Mower, a lovelorn young reaper meditating on the cruelty of his neglectful beloved, Juliana. This stanza from "The Mower's Song" is representative:

My mind was once the true survey
Of all these meadows fresh and gay,
And in the greenness of the grass
Did see its hopes as in a glass;
When JULIANA came, and she,
What I do to the grass, does to my thoughts and me.

(ll. 1–6)

Far from a literal meditation on biomass, the grass and Damon's mowing serves as a metaphor for unrequited love, a popular subject at court. Eventually the grim reaper and death will populate Damon's sad musings. Rather than being celebrated for its life-giving energy, here the grass-to-hay labor is a killing spree. Damon is literally the reaper, and figuratively his own quarry, the lopped grass. Juliana is the scythe. Marvell does not figure her as the sun that makes the grass grow, the rain that makes it supple and shiny, the breeze that causes it to dance. She shows Damon his mortality by severing his vitality—perhaps a very specific part of it.

The lovelorn lad appears again in Marvell's longer poem "Damon the Mower," this time with more details concerning his role in the rural network. The tetrameter couplets lend a light, ballad-like tempo to the verse that, by design, joins pairs of images into happy unions, like the dew on the daffodils:

> I am the Mower Damon, known
> Through all the meadows I have mown.
> On me the morn her dew distills
> Before her darling daffodils.
> And, if at noon my toil me heat,
> The sun himself licks off my sweat.
> While, going home, the evening sweet
> In cowslip-water bathes my feet.
>
> What, though the piping shepherd stock
> The plains with an unnumbered flock,
> This scythe of mine discovers wide
> More ground than all his sheep do hide.
> With this the golden fleece I shear
> Of all these closes every year.
> And though in wool more poor than they,
> Yet am I richer far in hay....
>
> How happy might I still have mowed,
> Had not Love here his thistles sowed!
> But now I all the day complain,
> Joining my labour to my pain;
> And with my scythe cut down the grass,
> Yet still my grief is where it was:
> But, when the iron blunter grows,
> Sighing, I whet my scythe and woes.

(ll. 41–56, 65–72)

Damon is an active figure set against the shepherd's idle idyll, a worker who relishes the natural relief of morning dew, midday sun, and a footbath of flower juice. He covers a wider terrain than the shepherd to cut his "golden fleece" of hay, a commodity that he presumably sells or trades for necessities. Without his work, the sheep and other quadrupeds would not eat in winter; so, though he is a figure of late summer, Damon's reaping presages the dark months of the year. The transition—from a poem that describes actual work to a poem that fixates on the emotive figures of the lovesick—is as inevitable to English pastoral poetry as the pouncing couplets. The thistles that dull his blade also barb his heart. Accepting that this kind of poetry was not primarily written to describe rural work or the collection of biomass to feed domesticated animals, early modern pastoral works nevertheless preserve an ideal version of a muscular way of life that has been annihilated by fossil-fuel-driven machines.

In his poetry, Marvell's contemporary, John Dryden, bypassed the tradition of the lovelorn rustic and went straight to the harvest party. Dryden's *Song* within the masque *King Arthur* flies on the voice of Comus, God of the bender and the hangover, using the hard labor of haymaking as an occasion for a party:

> Comus: Your hay it is mow'd, and your corn is reap'd;
> Your barns will be full, and your hovels heap'd:
>> Come, my boys, come;
>> Come, my boys, come;
> And merrily roar out Harvest Home.

(ll. 1–5)

These farm boys are ready for the call-to-goblets. They echo Comus in their chorus: "Come, my boys, come; / Come, my boys, come; / And merrily roar out Harvest Home." With winter food for human and animal secured, the last fine days of the season may be spent in celebration of nature's bounty, an overflowing cornucopia trampled by drunken Bacchus. The pure animal pleasure of rest after hard work is an emotional acme of biomass literature, a state of somatic happiness that industrial humans rarely access. When we do feel this bliss, it is most often caused by the elective labor of exercise rather than the life-sustaining work of the harvest. That work is now mostly performed by fossil fuels in machines, so we turn to treadmills as a muscle-working but energetically counterproductive medium. Our dwellings are filled with two types of machines: the ones that use fossil fuels to "save" labor, and the ones that use fossil fuels to absorb it—a paradox stored side by side.

As the eighteenth century passed and urban industry drew waves of rural people into the cities, the depopulated country did not offer an easy alternative. On the contrary, bad weather caused a series of bad harvests in Europe in the 1780s, including the near-famine in France that fomented the great revolution of 1789. In England, the price of wheat was 66 percent higher in 1783 than it had been in 1780 (Archer 30). The British Romantic poet William Blake deploys the pastoral to evoke the Christian tradition of Christ the Good Shepherd, with the church as his flock, safe and secure, a mode of innocence to be both celebrated and questioned, since it coexisted with a world of rampant victimization. In

Blake's "The Shepherd" from his *Songs of Innocence*, the pleasure is simple and absolute:

How sweet is the Shepherd's sweet lot!
From the morn to the evening he strays;
He shall follow his sheep all the day,
And his tongue shall be filled with praise.

For he hears the lamb's innocent call,
And he hears the ewe's tender reply;
He is watchful while they are in peace,
For they know when their Shepherd is nigh.

(ll. 1–8)

Something tastes sour in the initial repetition of "sweet," implying that the shepherd's innocence in this short ballad is perilously idealistic. Blake lived at a time when the shift from biomass to fossil-fueled society was accelerating, and the lifestyles of rural and urban dwellers were drawn into sharper relief. The shepherd is watchful, hear-ful, and he strays all the day *sans ennui*. The language of the pasture is the mother ewe's babbling love speak, translated into eight gentle rhyme-embraced lines. The pragmatic business of grazing is overshadowed by the filial lovefest, for work itself is easy pleasure, a passive receptiveness to the bounty of the sun. Blake would revisit these themes in his *Songs of Experience*, but this *Innocence* poem has no clear counterpart as many of the others do (for example, "The Lamb" in *Innocence* is matched with "The Tyger" in *Experience*). One possible complement to "The Shepherd" is "The Chimney-Sweeper," a protest lyric about the oppression of young boys in London that will appear in our readings on coal. The two poems form the ultimate critical dialectic between country and city in Blake's Romantic–industrial poetics.

In "I Heard an Angel," Blake deepens his dialectic between innocence and experience by embodying these states in angels and devils. Blake does not believe in traditional oppositions of good and evil propounded by institutional religion. As explored richly in *The Marriage of Heaven and Hell*, this shorter lyric shows how devilish attributes create a more complex and ultimately richer human experience. Morality plays out on the stage of the pasture:

I heard an Angel singing
When the day was springing
Mercy Pity Peace
Is the worlds release

Thus he sung all day
Over the new mown hay
Till the sun went down
And haycocks looked brown

I heard a Devil curse
Over the heath & the furze
Mercy could be no more
If there was nobody poor

And pity no more could be
If all were as happy as we
At his curse the sun went down
And the heavens gave a frown

Down pour'd the heavy rain
Over the new reap'd grain
And Miseries increase
Is Mercy Pity Peace

(ll. 1–20)

The angelic world kills those attributes that we prize as the best of human moral potential: "Mercy Pity Peace." The pastoral idyll is the one-dimensional stage of the perfect, paradisiacal condition. When the Devil's curse brings on the rain to spoil the hay, suffering, poverty, and penury excite the mercy that had lain fallow in the bronzed harvest earth. Blake's blasting curse slashes an impossible ideal and deepens the human (and animal) experience in the character-forming hard times to come. Blake apes the pastoral idyll in "The Shepherd," and he openly decries it in "I Heard an Angel."

Romantic poets are stigmatized for idealizing nature in flowery verse, but Blake's poetry shows how a transitional wind blows in several directions. Rural writers of the period like John Clare made a career of lamenting the changes to the countryside, and his nostalgic view is more traditionally pastoral. In his ode "Summer Images," he streams through a series of bright and mazy vales of rural pleasure:

Or thread the sunny valley laced with streams,
        Or forests rude, and the o'ershadow'd brims
Of simple ponds, where idle shepherd dreams,
        Stretching his listless limbs;
Or trace hay-scented meadows, smooth and long,

> Where joy's wild impulse swims
> In one continued song.
>
> (ll. 92–8)

Lovely, if a tad clichéd. It unites the components of an ideal evolutionary landscape: valley, forests, ponds, meadows. E. O. Wilson's *Biophilia* (1986) explores the idea of nature love as an evolutionary adaptation that informs our landscape aesthetics. The shepherd's idleness and "joy" are in pathos misplaced in nature rather than in a happy human perceiving nature. It is the poet's joy of imagination. At a time of mass land transformation, human migration to cities, and industrial pollution, Clare's verse shows the power of literature to lift the reader out of messy modernity by way of the nostalgic imagination. There is plenty of fragrant hay in store for the winter that, in this classical idyll, will never come. Clare's poems are full of these rhetorical idealisms; but the rural perfection he captures is also rhetorical, arguing for a dying way of life and an endangered landscape. Clare once said "I live here among the ignorant like a lost man," indicating that his experience of the pastoral was textured with melancholy, isolation, and long periods of mental instability (Bate 237). He was in and out of asylums for nearly thirty years in his later life, and died in 1864.

The 1840s–50s had witnessed particularly aggressive enclosure acts that advanced the centuries-long process of privatizing the English countryside to make for more homogenized, profitable agriculture that fed the urban industrial hordes, many of whom were compelled to the city because they were driven from subsistence on the old commons land. Enclosure paved the way for higher rents on tenants, payable to landlords who claimed to improve the land by crossing it with fences. With machines also displacing rural workers at a greater scale every year, Clare's lifetime was one in which hay and corn and the animal labor they supported greatly diminished from their former role as the nucleus of human life. Ironically, the modern "English Countryside" of crisscrossed fields is a product of the enclosure acts—so our perception of the pastoral is built upon the very structures that destroyed the classic pastoral of the commons.

Jane Austen riffs on the pastoral to poke fun at silly characters who put on airs with their ecstasies over picturesque prospects. There is not much real work in her novels, if one excludes the ruinous working-over of social cliques. *Mansfield Park*, though, glimpses at country work, if only as a setting to exhibit the self-absorption of the dangerously alluring Mary Crawford. Austen sagely chooses a pastoral instrument, a harp, that Mary has bought and requires fetching to her house. She sees plenty of horses and carts in the country—why can she not engage one to do her bidding? Her cluelessness is on display to the hero Edmund:

> "I am to have it to-morrow; but how do you think it is to be conveyed? Not by a wagon or cart: oh no! nothing of that kind could be hired in the village. I might as well have asked for porters and a handbarrow."
>
> "You would find it difficult, I dare say, just now, in the middle of a very late hay harvest, to hire a horse and cart?"
>
> "I was astonished to find what a piece of work was made of it! To want a horse and cart in the country seemed impossible, so I told my maid to speak for one directly; and as I cannot look out of my dressing-closet without seeing one farmyard, nor walk in the shrubbery without passing another, I thought it would be only ask and have, and was rather grieved that I could not give the advantage to all. Guess my surprise, when I found that I had been asking the most unreasonable, most impossible thing in the world; had offended all the farmers, all the labourers, all the hay in the parish! As for Dr. Grant's bailiff, I believe I had better keep out of *his* way; and my brother-in-law himself, who is all kindness in general, looked rather black upon me when he found what I had been at."
>
> "You could not be expected to have thought on the subject before; but when you *do* think of it, you must see the importance of getting in the grass. The hire of a cart at any time might not be so easy as you suppose: our farmers are not in the habit of letting them out; but, in harvest, it must be quite out of their power to spare a horse." (46–7)

Mary mistakes the real work of country laborers for the picture show of the classical pastoral that she thinks she sees every time she looks out of her self-ornamenting "dressing-closet." In the process of emerging from this harp-plucking dream, she offends everyone who labors to survive the coming winter. Not a single rattletrap cart, let alone an able horse, can be spared when there is a tight timeline on airing and storing the winter fodder. It is not a matter of money, it is one of survival. She justifies herself by noting the difference between city and country culture: "I shall understand all your ways in time, but, coming down with the true London maxim, that everything is to be got with money, I was a little embarrassed at first by the sturdy independence of your country customs." The embarrassment is not for herself; it is a kind of disdainful pity for the farmers' lack of deference. Since neither her charms nor her money can persuade the country people to meet her demands, she calls in a favor to her brother Henry, who plans to "fetch it in his barouche." Some horses are fueled to work the harvest; others dutifully drag harps to distant ladies.

Where classical and early modern writers deployed the pastoral to record ritual and to entertain the privileged, Romantic writers and their descendants were faced with a more tortured task of challenging the great fuel shift into a fossil-driven world. If one of the key aims of this study is to identify when the consumption of fuels became reflexive, that is, recognized in art as a transition from one ontology to the next, the Romantics and Victorians are ground zero for overt representations of a new world order. Machines would arrive on the farm and transform the nature of rural labor by the mid-1800s when, for example, Thomas Hardy's Tess is shaken to pieces on a coal-fired corn rick.

Once the harvest was in, the Bacchanalia would follow. As Thomas Hardy captures in a short scene in *Tess of the D'Urburvilles*, biomass fills the air of the dance with a "floating, fusty *debris* of peat and hay, mixed with the perspirations and warmth of the dancers, and forming together a sort of vegeto-human pollen" (68). The workers are shellacked in their labor, a coating of biomass that translates by metaphor to the earthy spirit of the people. The "fusty" air is literally inspired by the dancers, who outpace "the muted fiddles [that] feebly pushed their notes, in marked contrast to the spirit with which the measure was trodden out. They coughed as they danced, and laughed as they coughed" (68).

In strong contrast to the air pollution of coal—an atmosphere familiar to the Victorian Hardy—this rural suffocating air is joyous. It is an incantation of earth and countryside, a hazy veil that obscures clear vision and enhances the romantic effect: "Of the rushing couples there could barely be discerned more than the high lights—the indistinctness shaping them to satyrs clasping nymphs—a multiplicity of Pans whirling a multiplicity of Syrinxes; Lotis attempting to elude Priapus, and always failing" (68). Hardy riffs on the erotica of the classical pastoral with his mythological revelers—an attempt to briefly reanimate the nostalgic genre—before history squelches it for industrialization.

Before long, Tess and her friends must return to the coal-powered hay trusser, where their labor is chained to a machine pace and they can no longer share conversation over the racket. Tess's mechanized farm labor shows how the incursion of coal-fired steam power fatally disrupted the traditional methods of muscle labor and the evolved human experience of slow, steady work, along with the sensory redolence of the biomass-fueled farm. A new scheme of erotic joy, or at least release, would soon come into orbit around the urban environments built from the energy of coal and oil, starring the steam and thrust of the factory engine.

Half an industrial century later, Robert Frost shows that haymaking is endangered labor in America in his long poem "The Death of the Hired Man," written in 1905–6. It is winter, and a farming couple debate whether to shelter a former worker who left in the middle of hay cutting season and has now returned in a desperate state. The history of this man, Silas, emerges as the couple recalls his feud with another worker, a younger school boy enamored with his studies, particularly Latin. Using the two haymakers, Frost draws out the generational tension between the value of labor and that of abstract knowledge. Silas seems to be filled with regret for his former invective against Harold's learning, but he keeps his pride of work when he imagines a new encounter in the coming summer season: "He thinks if he could teach him [haymaking], he'd be / Some good perhaps to someone in the world. / He hates to see a boy the fool of books" (ll. 96–8).

As preordained in the title, Silas is flashing upon a future he will never see. We imagine Harold has gone on with his education and the upward mobility it enables, toward professional work and urban living supported by an industrial economy. His departure leaves haymaking to a dying generation doting on the labor practices of the diminished rural sphere. Silas is a cryptic character, out of grace with family and friends and dying a beggar's death. He is no farm-fresh paragon to set against the morally ambiguous white-collar classes: Frost's conflict is not bucolic versus bourgeois. The death of Silas is a symbolic extinction of preindustrial rural knowledge and, with it, any future interest in such knowledge, as the twentieth century turns toward fossil-fueled large-scale agriculture misnamed the Green Revolution. Silas and Harold had cut grass by hand and piled the hay upon a horse-drawn cart. Haymaking will soon be done by machines, and Harold will not be in the driver's seat. He will sit at a distant desk.

In Sarah Orne Jewett's poem of 1882, "A Country Boy in Winter," the boy belongs to Silas's generation, torn between rural and urban, pastoral and industrial. The long dark seasons of quiet time would be insufferable to fashionable people used to constant stimulation, but the boy finds ample company and cheer in the working animals he is responsible for. His care is the work of maintenance, but also of love, a union known as husbandry:

> I like to hear the old horse neigh
>   Just as I come in sight,
> The oxen poke me with their horns
>   To get their hay at night.
> Somehow the creatures seem like friends,
>   And like to see me come.
> Some fellows talk about New York,
>   But I shall stay at home.

(ll. 41–8)

The fellows talking of New York choose an urban life, which depends on the work of faraway farms and mines and factories and fisheries. The city itself, its art and consumerism and parties, is the luxurious enclave of those who exploit the productivity of other people in other places. By contrast, the affection of the old horse and the oxen is pleasure gained from embeddedness, an evocative straw-and-urine version of life that masses of young people were leaving behind, but that this country boy continues to cherish. Like Clare's plaintive ode to the bucolic, Jewett's more realist poem preserves the old world order as America careened into the fossil future.

Frost's sonnet "Mowing," written about ten years after "The Death of the Hired Man," is immersed in the sensuous task of manual mowing. It preserves an old mode of existence: the mental and physical feel of cutting grass with a scythe, feeling the variable elements, observing the other dwellers in the meadow. Frost critiques the classical pastoral while at the same time preserving some of its more appealing traits: quiet thought, observation, and naturally good company:

> There was never a sound beside the wood but one,
> And that was my long scythe whispering to the ground.
> What was it it whispered? I knew not well myself;
> Perhaps it was something about the heat of the sun,
> Something, perhaps, about the lack of sound—
> And that was why it whispered and did not speak.
> It was no dream of the gift of idle hours,
> Or easy gold at the hand of fay or elf:
> Anything more than the truth would have seemed too weak
> To the earnest love that laid the swale in rows,
> Not without feeble-pointed spikes of flowers
> (Pale orchises), and scared a bright green snake.
> The fact is the sweetest dream that labor knows.
> My long scythe whispered and left the hay to make.
>
> (ll. 1–14)

The mower is "beside the wood" in the pasture, a productive space that yields the necessary hay for fodder, silent but for the sound of the scythe's slice across the hay grass. The mower hears a convivial conversation between blade and blade, perhaps about the original source of life, the all-animating sun whose light has been chemically translated into biomass. Heavy, productive biomass: enough to make a "swale" from the reaping, a landform of its own laid out in rows to dry, again by the sun's energy. The mower refuses to countenance pastoral exaggeration; he has no interest in "more than the truth," or a life involving "idle hours" and "easy gold," peopled by fairies and fantasies. These abstractions are a literary failure to convey the real love of hard work and the actual beauties of "pale orchises" and "a bright green snake" that dwell in this world. The mower kills and scatters them, too, but not without appreciating them first. The hay meadow is a habitat, a polyculture in which the weeds are sexy orchids sheltering suggestive snakes. From the mundane drudgery of the reaping, Frost has created a whispering, pulsing, productive and reproductive world without a single reversion to a half-clothed Dryad. And the mower has laid down the year's harvest to cure, stored for the working animals behind the scene.

The aesthetic pleasure of Frost's sonnet lulls the reader into a pastoral reverie, the kind of space that Keats so deliciously crafted in his ode "To Autumn." But though the season of reaping may be "close bosom-friend of the maturing sun," sumptuous literary visions of reaping contain auguries of deprivation and death as well. Jim Crace's novel *Harvest* (2013) profiles a remote English village sometime in the long era of enclosure, the twelfth to the nineteenth century, when peasants were driven off the common lands so that private owners could pursue profitable industries like wool making and monoculture commodity farming.

Crace describes the collective synergy of feudal-style subsistence, where "the broadest shoulders swing their sickles and their scythes at the brimming cliffs of stalk; hares, partridges and sparrows flee before the blades; our wives and daughters bundle up and bind the sheaves, though not too carefully—they work on the principle of ten for the commons and one for the gleaning: our creaking fathers make the lines of stooks; the dun begins to dry what we harvested. Our work is consecrated by the sun" (6). Each individual has a place dictated by age or sex, but the places are equally important to survival. This warm-bellied passage does not lull the reader into innocent nostalgia for simpler times, however: Crace's characters are skinny, sinewy, scarred, and usually cold. The narrator evaluates the year's harvest: "I plunge my good hand into a half-filled sack ... I've known better harvests, years when the barleycorn was fat and milky ... And I've known hungry years when yields were fibrous and parched, and we survived the winter on dry bones. Today the grains are good enough, but only good enough. We will not starve; we will not fatten either" (72–3).

The narrator Walter Thirsk is a transplant from the city who might recognize an ode like Keats's as a city man's poem, since the hard work of harvest is impossibly performed by drowsy goddesses rather than straining muscles. Self-conscious of being an outsider, Thirsk wants to validate his presence by performing the physical effort that pastoral dwelling demands. That world is endangered, however, by enclosure that would substitute one form of pastoral, the plowman's agriculture, for another: the shepherd's ovine pastoral. The latter is the classical, idealized, and often urban-written version of the pastoral; the former is a genre of georgic that commits itself to the hard work of agriculture. The defeated local master, Kent, resists his more powerful cousin's happy vision: "The commons will be cleared and privately enclosed. You're pasture now. These lands are grass. We'll never need another plow" (91). Kent surveys his farm land before it is appropriated, and marks the old way dying for the new to be born: "This land has always been so much older than ourselves ... not anymore" (174). When the commons are cleared, the land is reborn as alien: "That drystone wall, put up before our grandpa's time and now breeched in a hundred places, will be ... replaced either with an upstart thorn or with some plain fence, beyond which the flocks will chomp back on the past until there is no trace of it. We'll look across these fields and say, 'This land is so much younger than ourselves' " (175).

Kent imagines a temporality here between old and young, where agriculture is the old land, and grazing is the young land, as perceived by the stewards who have seen it evolve through time. His view suggests a return to the innocence of the shepherding regime, where the land itself is young. These two faces of the pastoral, grueling subsistence and leisurely flocking, preview the dualism of fossil fuels, which created a world divided between grinding factory labor and the flighty indolence of automation and leisure-giving wealth. In both cases, the fuel dualism is borne on a social hierarchy that first divides workers from masters, then producers from consumers. However, this parallel between biomass and fossil fuels is also perpendicular: with biomass, the peasant's labor is arduous because the system's energy is so meagre; with fossil fuels, the proletariat's labor is arduous because the system's energy is tremendous. The factory worker has to sprint to keep up with the steam-driven line.

The pastoral is often filled with vague, utopian locales and stock characters. It can, however, become notably particularized. For example, James Whitcombe Riley's late 1800s doggerel, "Raggedy Man," evokes rural Indiana and the Hoosier dialect, as it reveals a boy's fascination with the hobo who worked the farm:

O the Raggedy Man! He works fer Pa;
An' he's the goodest man ever you saw!
He comes to our house every day,
An' waters the horses, an' feeds 'em hay;
An' he opens the shed—an' we all ist laugh

When he drives out our little old wobble-ly calf;
An' nen—ef our hired girl says he can—
He milks the cow fer 'Lizabuth Ann.—
    Ain't he a' awful good Raggedy Man?
      Raggedy! Raggedy! Raggedy Man!

W'y, The Raggedy Man—he's ist so good,
He splits the kindlin' an' chops the wood;
An' nen he spades in our garden, too,
An' does most things 'at *boys* can't do.—
He clumbed clean up in our big tree
An' shooked a' apple down fer me—
An' 'nother 'n', too, fer 'Lizabuth Ann—
An' 'nother 'n', too, fer The Raggedy Man—
    Ain't he a' awful kind Raggedy Man?
      Raggedy! Raggedy! Raggedy Man!

                         (ll. 1–20)

The exuberant cadence, its hollering and repetition, mimics the bellows of a farm boy. Raggedy Man fascinates because his uproarious bedraggled looks belie his competence; he does things "boys can't do" but might wish to, such as climb high in a tree to shake down apples. He is the lowest on the social rung—he needs to ask permission even from the "hired girl"—but in his way he's a prince of the pastoral, master of a series of can-do verbs: works, waters, drives, milks, spades, climbs, shakes. He's goofy, spontaneous, energetic, and ready, a farm-clown counterpoint to the boy's dour father. Riley's verse became part of popular culture when the creator of the Raggedy Ann and Andy dolls, first sold in 1915 to accompany a series of illustrated books, named them after Riley's character.

In "The Black-Faced Sheep," contemporary poet Donald Hall explores the enduring ties between humans and the animals we keep—not as a relationship of subordination or dependency, but as kinship in an evolutionary journey. As with the earlier writers, the pastoral scene is replete and redolent with life-sustaining grass and grain:

I forked the brambly hay down to you
in nineteen-fifty. I delved my hands deep
in the winter grass of your hair.

When the shearer cut to your nakedness in April
and you dropped black eyes in shame,

hiding in barnyard corners, unable to hide,
I brought grain to raise your spirits.

(ll. 24–30)

Hall's empathy for the sheep's grief invokes an existential kinship we hold with animals long involved in our survival. Sheep were among the first animals to be domesticated in the Fertile Crescent and northern Africa, and their wool and milk were essential to the development of stable settlements. Hall sweeps from today back to the origins of all pastoral literature:

Ten thousand years
wound us through pasture and hayfield together,
threads of us woven
together, three hundred generations
from Africa's hills to New Hampshire's.

(ll. 31–5)

The figurative fabric is woven of real hay and the wool it eventually produces, linking the global landscapes of animals at pasture through deep time. The thread of human–animal symbiosis ties to our traditions of husbandry, beginning long before humans had the ability to recall or record. Deep memory is also scripted into our genes: humans are selected by their ability to grow and cure ample fodder and care for animals, and sheep are selected to evolve into docile, noble, wooly mammals able to subsist on biomass we can't digest. The evolutionary ecology of animal domestication is an intimate marriage of equals who have wandered together across the globe. Industrial animal husbandry makes few ethical allowances for this biocultural codependence, but poetry takes up the project of recuperating it in the preindustrial pastoral.

With Virgil, we began looking at the origins of the pastoral as a study of rural human ecology, with the landscape as an indicator of the health and fortunes of the dependent shepherds. Virgil used the classical idyll not only to texture his poems, but also to boldly prophesy apocalypse—environmental disaster—as a drastic event that could occur in any earthly Elysium. Apocalypse has hardly ever gone out of fashion in literature. Readers have a macabre fascination with the nature of our own demise; we attribute the ruckus to mischance or retribution, and enjoy the spectacle of an ordered world devolved to chaos. The German writer Marlen Haushofer's contemplative, anxious novel *The Wall* (1968) flavors the pastoral with postmodern nuclear-age angst. It describes a postapocalyptic world of individual subsistence on preindustrial biomass. Our evolutionary conditions have been mostly erased just two hundred years after the advent of industrial civilization. Haushofer reawakens them.

The wall itself is an anomaly that Haushofer does not explain. It is simply a clear, impervious membrane that one day surrounds a woman visiting a hunting lodge in the Alps. Everything outside a few hundred walled-in acres is frozen in time, petrified, suggesting nuclear winter. The novel focuses on the sphere of the living, using an epistolary narrative to tell the woman's tale of survival and her life shared with animals: a dog, cats, two cows, deer, and a fox. The woman— unnamed, so I'll call her Woman—grows from pudgy, dissatisfied, and middle-aged, to lean, purposeful, reverent, and despairing. Autonomous, increasingly competent, contemplative, propertied, she is a female Robinson Crusoe; she is Eve alone in a garden of strenuous labor.

Woman wanders her world looking for answers, and comes upon a dairy cow in a barn who she names Bella. This valuable animal is a blessing for its milk, but it is also a burden due to the care required: the cow is, on one hand, her property; on the other, her jailor. But the responsibility for Bella and her dog Lynx keeps her from attempting suicide amid her crushing loneliness. To maintain Bella, she scavenges for stored hay in the surrounding buildings, repurposes the car garage as a hay loft (a fuel reversal), and grazes the cow in the forest meadow. She learns by error and effort how to milk a cow manually, which results in a kind of conversation: "Sometimes she turned her great head towards me as if she were watching my efforts with amusement, but she stood there quietly and never kicked at me; she was friendly, often even a little bumptious" (29).

Through that first summer Woman gains strength and wherewithal; she plants potatoes and beans, learns to hunt deer, and migrates between the hunting lodge in the valley and the cottage in the high meadow. One essential labor is collecting and gathering timber to keep warm in winter. Splitting wood is difficult and dangerous—she cuts a menacing gash in her knee that aches when the weather changes. Stored wood is literal and emotional security, and she works herself to exhaustion to lay in her supply.

She learns, though, that her single hardest labor is haymaking. It takes three weeks to scythe the meadow by hand, to dry and cure it in the barn. By August, "I was so exhausted that I sat down in the meadow and wept. I was overcome by a wave of despair, and for the first time I understood quite clearly the hard blow that had hit me. I don't know what would have happened if responsibility for my animals hadn't forced me at least to do the most necessary things" (65–6). The "blow" that hit her is the shock of physical limitations in a biomass world—her fossil-fueled body screams as it adjusts to the older regime. By the next summer's harvest, she does better: "By seven o'clock I was in the meadow by the stream, swinging the scythe ... I scythed for three hours, and I found it went better than I'd thought it would ... better than the previous year, when I had touched the scythe for the first time in twenty years and hadn't yet grown accustomed to hard work" (171). Physical pleasure translates to aesthetic immersion: "The sun cast its full brilliance on the slope. The fresh-cut swathes of hay already lay wilted and dull ... The meadow was one great hum of startled insects.... I abandoned myself entirely to the buzzing, hot stillness" (171). After more labor, enduring the elements and the fears of a lonely imagination, she finishes the year's greatest task, the haymaking. She compares her labor to motherhood: "I couldn't remember having felt so much satisfaction since my children were little. Back then, after the strain of a long day, when the toys were cleared away and the children lay in their beds, bathed and dried, back then I'd been happy" (177). Her haymaking is a new form of motherhood, the meadow grass to be cleared and stored like her children's toys. In a biomass realm, working outdoors is embraced within a domestic sphere. Its burden is the work: constant energy staves off entropy; its reward is the milk, the meat, the offspring.

Yet she perceives Bella the cow not as her daughter, but as her mother: "When I combed Bella I sometimes told her how important she was to us all. She looked at me with moist eyes, and tried to lick my face.... Here she stood, gleaming and brown, warm and relaxed, our big, gentle, nourishing mother. I could only show my gratitude by taking good care of her." (163–4). Bella is a mother, she gives birth to a calf, but she also mothers all the survivors with her matronly heifer-ness, her ability to nourish them by turning the meadow into cream. The cream is a delicacy when Bella grazes in swaths of sweet wild herbs.

Woman finds her most comforting habitat in the high meadows, the place where the rhythms of her animals and the fecund wildness all around her comes into concert with her cognitive state. She rhapsodizes:

> What I liked doing best was simply looking out over the meadow. It was always in gentle motion, even if I thought the wind was still. An endless, gentle ripple emanated peace and sweet fragrance. Lavender grew here. Alpine roses, cat's foot, wild thyme and a host of herbs whose names I didn't know; which smelt just as good as thyme, but different. ... The meadow slowly went to sleep, the stars came out, and later the moon rose high and bathed the meadow in its cold light. I waited for those hours all day, filled with a secret impatience.... The great game of the sun, moon, and stars seemed to be working out, and that hadn't been invented by humans. (183–4)

She lives the life of the classical shepherd, with long days beneath the changing sky filled with meaningful interactions with the land and its animals, so different from the drudgery and ennui of her now-extinct industrial life. By dropping out and tuning in, Woman discovers a larger sphere of existence, beyond the world of competitive consumption ushered by fossil fuels. Now attuned to the cosmic ecstasy of life on earth, she pays closer attention to the details of her surroundings, to species known and unknown, and what they inspire in her awakened senses. She can now track and appreciate the intimate relationship between biomass, cow, and human—an intimacy that is missing from the anonymous production and distribution of goods in the modern food system.

The apocalypse of the wall yields a postindustrial pastoral idyll, though Haushofer is careful not to romanticize hard physical labor, the revulsion of isolation, and terror of the unknowable future. Written at a time of recuperation after the Second World War's hideous industrial genocide, when the nuclear specter loomed over transnational relations, Haushofer's novel participates in the earth-loving fashion of the 1960s while resisting its sentimentalities. It displays the extremes of pleasure and pain inherent to a biomass ontology, where physical work provides a deep sense of dwelling while bowing the mower's back deeper year after year.

Literature has identified an array of unique qualities in rural laborers and their work, especially as this kind of work has diminished in importance during the industrial era. Pastoral writers turn what is literally the growing, gathering, and consumption of biomass into an entire ontological sphere of existence that is tuned into the seasons, the cycles of day and night, and the physical rhythms of a life yoked to gravity, topography, and weather conditions. Pastoral writing is affectionate and protective, innocent of corruption but experienced in hard work, rewarded by rest. It is poignantly opposed in space to cities and in time to modern industrialism, where fossil-fueled machines annihilate the physics and aesthetics of the ambient, hushed biomass world.

Postapocalyptic novels like James Howard Kunstler's *World Made by Hand* envision a world re-pastoralized, with heroes discovering their latent talents for handicraft, community barter and support, and vigilante-style resistance of outside threats. Kunstler's peak-oil politics inform his ideal of a post-fossil civilization that works hard to attain a modicum of stability and comfort. His specific vision, however, reverts not only to muscle power but also to stifling traditional gender roles with a kind of "men used to be real men" swagger, leaving women passive, domestic, dependent.

His nostalgic patriarchal vision raises the interesting question of whether fossil fuels have assisted in the liberation not only of slaves in the American south but also of women from the domestic confine, and whether a new post-fossil world has enough margin to permit the freedoms that came with a cheap-energy civilization (Bergthaller 430). Or, does that future world have so meagre an energy margin that no hand, whatever the sex and race, can be spared from hard work? Haushofer's novel declares that laboring females, human and bovine, can orchestrate everything. Kunstler's novel treats females more like livestock: useful assets for subsistence, but not equal participants in the novel's title: World Made by [Male] Hand.

A more spectacular version of fuel regime change takes place in Paolo Bacigalupi's *The Windup Girl*, where a climate-changed Bangkok of the 2200s is endangered by food shortages caused by corporate control of agriculture, widespread crop failures, and rising sea levels. Most of the heavy labor is performed by a genetically engineered form of elephant called the Megodont, a beast who galumphs in circles winding "kink springs," which can then be deployed as a kinetic motor in an array of tasks, most notably in the scooters that have replaced fossil-fueled automobiles. In *The Windup Girl*, the Megodonts pace crumbling highways; cities built for fossil-fueled trade swarm with draft animals and schooling bicycles.

Biomass energy flows through massive Megodont muscles into the kink springs, which relieve human muscles of the work they would otherwise have to perform through a direct biomass conversion. Since human food is so limited, the Megodont's ability to digest rough biomass again helps humans survive a hostile climate, not unlike the ruminants that permit herding societies to thrive on marginal lands where arable land for agriculture is limited. In this Bangkok, the government's trade industry has control over the remaining resources and their regulation, and these officials are the only ones permitted to use fossil fuels—in this case, diesel derived from coal—to power conventional automobiles. The raw power at work is so strange to the commoner, its effect is nauseating. In a world that moves at foot speed, the body exposed to diesel energy feels the sublime—a mixture of awe and terror:

> The exorbitant cost of turning this steel behemoth into acceleration. An extraordinary waste ... There is an ancient solidity to the thing, so heavy and massive—it might as well be a tank ... Claustrophobia swallows him ... The car accelerates, pressing Hock Seng into the leather seat. Outside the windows, ... crowds to sun-drenched skin and dusty draft animals and bicycles

like schools of fish. Eyes turn toward the car as it forges past. Mouths open wide
and silent as people shout and point at his passage. The speed of the machine is
appalling. (134)

This juxtaposition of biomass and fossil energy is temporal: though the two
forms coexist in this oppressive world, the limousine glass that divides the haves
from have-nots is also a barrier between past and present. We are accustomed to
thinking of biomass as past and oil as present, but in postapocalyptic visions, it
is oil in the past and biomass in the present. By reversing the coherent historical
evolution of fuel sources from low energy toward high energy, Bacigalupi and
Kunstler illuminate the chaos that ensues when societies are forced from energy-
dense to energy-sparse regimes. Energy becomes a scarce resource controlled
by those in power, rather than that enjoyed at will by the masses, and the
entire infrastructure organized around fossil fuels is ungracefully retrofitted to
accommodate the downgraded world order.

The pastoral is a powerful idea. It has long promoted Nature as a nurturing,
stable mother who provides refuge from the ugliness and abuse of city life, and
serves as an antipode the masculine forms of ecocultural representation—urban
and wilderness (A. Wilson 94). From its origins in working landscapes that fed
productive animals, these variegated grassy scenes have translated into sprawling
one-species grass lawns, especially in Europe and former colonies that inherited
European landscape aesthetics. The idle lawn designates dominion, leisure,
control, and wealth, which aristocrats have deployed for half a millennium to
distinguish their dwellings from those of their subordinates. The energy involved
includes not just the sun-infused grass but also the energy of countless laborers
who cultivate this unnatural space and, more recently, the petroleum products
that fertilize and mow millions of suburban lawns worldwide.

In postwar petro-America, white-collar work promoted mental work and
physical exertion waned, so men living in new suburbs often turned to intensive
"cultivation" of their plots, usually by wielding a lawnmower, while their wives
turned to flowery forms of gardening aimed at keeping up appearances (A.
Wilson 99). More on suburbs when we look into fossil fuels—but the thumbnail
portrays a widespread madness driven by the physio-cognitive dissonance
between our preindustrial evolution and our industrial condition.

Wild meadows remain grassy due to flooding, acidic soil, wildfire, or grazers.
Humans have innovated on the pastoral by artificially selecting animal species to
transform grass into milk, meat, wool, and labor. More recently, we have inserted
alien forms of fuel, especially petroleum, into the pastoral scene. With such

outlandish alterations, the pastoral is no longer a celebration of our subsistence upon nature; it has become the mental playground of a leisure class, the profit base for industries ironically dependent upon machines and petroleum. The feminine origins of the pastoral helped justify masculine exertion and dominance of the landscape using machines (A. Wilson 96). The use of the pastoral as a mental playground has classical origins, but its usefulness as an escape only intensified with the alienations of industrialism. The pastoral profit motive, the deployment of the trope to sell grass seed and fertilizer, to recruit buyers to new subdivisions, to rebrand industrial milk and meat as "natural"—that species of pastoral is a bastard scion of fossil fuels.

II The city: Crosstown traffic

Cities are animal rookeries. Living densely, breathing the atmosphere of bodies steeped in the elements, the urbanite from the city-states of antiquity to the megacities of the early industrial era had extensive contact with animals. Biomass-eating ungulates served the majority of tasks in transportation and food. Authors of the urban dung heap, their prevalence created the living conditions—sight, smell, and sound—that characterized the preindustrial urban sphere. Rural animals fit much better into the nutrient cycle than urban ones, since their end products could reunite with the soil directly. The dung of grass eaters could also be dried in patties and burned as fire fuel, a valuable alternative to scarce timber. However, to feed beasts of burden and open arable land for tillage, forests were frequently converted to pasture and farmland. In the seventeenth century, the trees in St John's Wood near horse-powered London were felled to make room for grasslands that produced two or three harvests of hay annually. In just two years in the 1820s, toll booths collected on twenty-six thousand loads of hay and straw entering the city (Velten 45). Fodder sold at the London locales of Haymarket, Smithfield, and Paddington steadily increased in supply until the peak demand around the turn of the twentieth century, when as many as seven hundred thousand working horses pulled the city's wheels (Velten 43).

Jonathan Swift amusingly describes the London dust in his 1710 poem "Description of a City Shower":

Not yet the dust had shunned the unequal strife,
But, aided by the wind, fought still for life,

Figure 2.2 Camille Pissarro, *Peasants Carrying Hay*, 1900. Image courtesy of the National Gallery of Art. Public domain.

> And wafted with its foe by violent gust,
> 'Twas doubtful which was rain and which was dust.
> Ah! where must needy poet seek for aid,
> When dust and rain at once his coat invade?
> Sole coat, where dust cemented by the rain
> Erects the nap, and leaves a cloudy stain.

(ll. 23–30)

The beleaguered poet shares a fate common to all inhabitants: Tories and Whigs, lowly seamstresses and gallant beaus. The rain mixes with smoke pollution and the animals' euphemistic "dust" to make plaster of Paris—well, of London—that shellacs them with the by-products of London's chimneys and working animals. To use animals, one must live intimately with them and endure the end effects of their rumination. A cow turning cellulose into sirloin may seem magical, but it is in fact earthy—it produces "earth." Other parts of the human–animal alliance come out in this wash, too. The repugnant climax of the city shower is a flood of dismembered bodies:

> Filths of all hues and odors seem to tell
> What street they sailed from, by their sight and smell.
> They, as each torrent drives with rapid force,
> From Smithfield or St. 'Pulchre's shape their course,
> And in huge confluence joined at Snow Hill ridge,
> Fall from the conduit prone to Holborn Bridge.
> Sweepings from butchers' stalls, dung, guts, and blood,
> Drowned puppies, stinking sprats, all drenched in mud,
> Dead cats, and turnip tops, come tumbling down the flood.

(ll. 55–63)

This amusing shit show reveals different quarters of the city by specialty: Smithfield is known for its cattle and meat market, St. Sepulchre's—opposite the infamous Newgate Prison—is the conjunction of high Anglicanism and low criminality. Into the Thames via Holborn flows the city's feces and cadavers. Humans cling together under the awnings while the lowering heavens sweep away the disgusting remnants of lives dependent on animal bodies. The poem's tone is one of comedic apocalypse: the dire spectacle of dirt and death in "the flood" literally washes the city of its sins, and after the end times comes a new beginning. The dung cycle renews.

The city is an ecosystem, too, and eighteenth-century Londoners welcomed rain and weather fronts to clear out the stalls and stables just as their nineteenth-century counterparts yearned for wind and rain to relieve the city of its coal dust. Rain has a marvelous power to absolve cities of their grimes. The pulse of water through the air and across the impermeable surfaces of the cityscape is a ready metaphor for renewal, one that Swift certainly employs here. But the metaphor is subordinated to the insistent adjective–noun listing of things literally washed into the Thames by the rain—stinking sprats, dead cats, and all. The rainy-day Thames is a toilet flushed.

Imagine the torrents of dung coursing down London streets, and you will cease to wonder that the horseless carriage was hailed as a clean alternative to the omni-sensory offenses of the rump. At the turn of the twentieth century in America, large cities like New York held a standing army of 120,000 horses, and even smaller cities like Rochester and Milwaukee—with several thousand horses each—were concerned that the annual pile of dung, 175 feet high and an acre in area, could breed sixteen billion disease-spreading flies (Tarr). What is a bit of carbon monoxide from a tailpipe compared to that sublime mountain of poo? Some urban manure was collected and sold as fertilizer, but this by-product of biomass-driven muscle power was mainly a dreaded reeking dusty nuisance that helped pave the way for cars.

So what if the cleaner spirit of a preindustrial age is captured in horse-driven cities? Thomas DeQuincey, in 1849's "The English Mail-Coach," uses the bodily effort of the horse to draw a philosophical distinction between horse-drawn coach and railway travel. The railway boosters boast of the velocity of the trains, the super-animal speeds and distances they attain. But DeQuincey, a loquacious narrator, argues that their chatter misses the point.

> Seated on the old mail-coach, we needed no evidence out of ourselves to indicate the velocity.... The vital experience of the glad animal sensibilities made doubts impossible on the question of our speed; we heard our speed, we saw it, we felt it as a thrilling; and this speed was not the product of blind insensate agencies, that had no sympathy to give, but was incarnated in the fiery eyeballs of the noblest amongst brutes, in his dilated nostril, spasmodic muscles, and thunder-beating hoofs. The sensibility of the horse, uttering itself in the maniac light of his eye, might be the last vibration of such a movement; the glory of Salamanca might be the first. But the intervening links that connected them, that spread the earthquake of battle into the eyeballs of the horse, were the heart of man and its electric thrillings—kindling in the rapture of the fiery strife, and then propagating its own tumults by contagious shouts and gestures to the heart of his servant the horse. (283–4)

The animus and emotive glory of being alive is transferred from the galloping horse to the thrilled rider, who, at eight mph, feels more alive than the rail rider at fifty. He celebrates the empathy between horse and human, feeling the fire and light of the body electric, an affinity that, to DeQuincey, is impossible to cultivate between human and machine.

> But now, on the new system of travelling, iron tubes and boilers have disconnected man's heart from the ministers of his locomotion. Nile nor Trafalgar has power to raise an extra bubble in a steam-kettle. The galvanic cycle is broken up for ever; man's imperial nature no longer sends itself forward through the electric sensibility of the horse; the inter-agencies are gone in the mode of communication between the horse and his master out of which grew so many aspects of sublimity under accidents of mists that hid, or sudden blazes that revealed, of mobs that agitated, or midnight solitudes that awed. Tidings fitted to convulse all nations must henceforwards travel by culinary process; and

the trumpet that once announced from afar the laurelled mail, heart-shaking when heard screaming on the wind and proclaiming itself through the darkness to every village or solitary house on its route, has now given way for ever to the pot-wallopings of the boiler.... . The gatherings of gazers about a laurelled mail had one centre, and acknowledged one sole interest. But the crowds attending at a railway station have as little unity as running water, and own as many centres as there are separate carriages in the train. (284–5)

The heart of a collective human purpose is "disconnected" from animals by the boiling pot of the steam locomotive. For DeQuincey, the fellowship between horse and rider is a sublime alliance in line with the high points of human empathy. Travel and delivery by horse offers irregular rhythms in a variable ratio of reward. It is not slotted into industrial scheme that shrinks the distance between London and Edinburgh into a numbered cell on a timetable. Likewise, the rider and the people at his destination have "one centre" and "one sole interest," whereas the crowds on the railway platform embody the anonymity and oblivion of a million self-interested atomies in continual dissolution. The horse has gravity; the locomotive, repulsion.

DeQuincey's florid style exaggerates to the point of eliding the similarities between horse and steam engine in the modernizing transportation system. By the nineteenth century, an improvement in roads through macadamizing meant that stagecoaches and mail coaches were on strict timetables that rivaled the efficiency of the train, and passengers in a coach were just as likely to ignore one another as those on a railway carriage. The "electric sensibility of the horse" was often neglected in preference for a practical and often cruel extraction of maximum labor from an animal that was easily replaced when exhausted. We look to DeQuincey for the spirit of the idea, more than its reality in the traffic of the early industrial era. His spirit is tried by his times as he bears witness to the new steam-driven world and its standards of efficiency applied across the board. This kind of writing, characteristic of the Romantics and Victorians, marks the time in which society became self-aware of the implications of its fuel regimes and, in so doing, entered the ideology of the Anthropocene.

III In depth: *Black Beauty*

If we want to take it straight from the horse's mouth, the voice of Black Beauty will give the best testimony of the nineteenth-century urban horse. Anna Sewell's novel has remained a perennial favorite since its publication in 1877. It reads as a form of slave narrative, where the horse's biography pulls the reader through a series of working situations, owners cruel and kind, companions separated by sales, and the long-term toll on health exacted from the body of the laborer. A work primarily concerned with animal welfare, *Black Beauty* pointedly protests specific forms of horse fashioning such as tail docking and the check rein that kept the heads of aristocrats' carriage horses unnaturally high.

I would like to examine three aspects of the novel that have received less attention: Black Beauty's relish of good food, his first-hand experience of the idyllic pastoral, and his musings on steam engines as a rival form of transportation. Turning the biomass of grass and grain into kinetic energy, our equine narrator views the great iron horse snuffing coal as an alien life form. But in the age of steam-driven long-range transportation, Black Beauty's body is too often mistaken for an iron machine.

He starts his story where many autobiographers do, in his infancy where his mother and her milk dominate his experience. That window of intimacy soon closes, for "as soon as I was old enough to eat grass my mother used to go out to work in the daytime, and come back in the evening" (1). Just like the human child of a working-class mother, Black Beauty admires his mother's energy and wisdom and has too little time to absorb her influence before he is sold into the laboring system. From then on, his evaluation of goodness depends upon decent treatment, nourishing food, and a healthy habitat with clean straw, fresh air, freedom of movement, and light. Sewell's rhetorical success depends upon the reader's empathy as they see that the pleasures of a horse mirror those of a human. He relishes "very nice oats," "very sweet apples," deep meadow grasses, and bran and boiled linseed mash (34). When he is recovering from overwork and abuse, oats play a large part in his mending, both physical and emotional. His companion Ginger shares an episode of abuse where fodder plays a major part in her recovery; kind food is inseparable from kind words: "The skin was so broken at the corners of my mouth that I could not eat the hay, the stalks hurt me. He looked closely at it, shook his head, and told the man to fetch a good bran mash and put some meal into it. How good that mash was! and so soft and healing to my mouth. He stood by all the time I was eating, stroking me and talking to the man. 'If a high-mettled creature like this,' said he, 'can't be broken by fair means, she will never be good for anything'" (34).

Even when physical mistreatment is not the problem, miserly stable keepers can ruin the health of the horse by withholding proper food. In one episode,

Black Beauty is owned by a kind but clueless man who is cheated by his hostler, who reserves expensive oats for his other business of fattening poultry and rabbits. "After awhile it seemed to me that my oats came very short; I had the beans, but bran was mixed with them instead of oats, of which there were very few; certainly not more than a quarter of what there should have been. In two or three weeks this began to tell upon my strength and spirits. The grass food, though very good, was not the thing to keep up my condition without corn" (152–3).

Black Beauty is stoic, but eventually his health fails and the owner's perceptive friend notices the problem: "'He is as warm and damp as a horse just come up from grass. I advise you to look into your stable a little more. I hate to be suspicious, and, thank heaven, I have no cause to be, for I can trust my men, present or absent; but there are mean scoundrels, wicked enough to rob a dumb beast of his food. You must look into it.'" (154). The owner's friend gives the kind of acute perceptiveness here that DeQuincey celebrates in his human-equine union of mind and momentum. We are capable stewards when we focus our attention on the well-being of the animal, and ennobled in that pursuit. When the cheating hostler is found out, he is sentenced to two months in prison, a rare victory for the rights of the horse. The penalty is measured anthropocentrically in proportion to the crime—sabotage of another man's property.

A working horse's life story has a different trajectory than the classic bildungsroman; it is plotted on a sad, downward slope as the horse loses strength and spirit through a succession of hard jobs. Black Beauty suffers injuries enough to bring down his value from a fine carriage horse to a low cart hauler, and this job in his older years nearly ruins him. After a bad fall in the city street, his self-interested owner chooses to elevate Beauty's resale value rather than shooting him, an option with minimal financial return for the meat and hide. So Skinner, out of avarice rather than kindness,

> gave orders that I should be well fed and cared for, and the stable man, happily for me, carried out the orders with a much better will than his master had in giving them. Ten days of perfect rest, plenty of good oats, hay, bran mashes, with boiled linseed mixed in them, did more to get up my condition than anything else could have done; those linseed mashes were delicious, and I began to think, after all, it might be better to live than go to the dogs. When the twelfth day after the accident came, I was taken to the sale, a few miles out of London. I felt that any change from my present place must be an improvement, so I held up my head, and hoped for the best. (250)

This investment in high-quality feed is a success for all involved: Beauty is resold at a good price to a kind country man, who turns him out to pasture to knit his knees for a few months before landing his last job as a carriage horse for leisurely ladies. The carefully regulated and diverse diet of the working horse—oats, bran, linseed, apples, in addition to grasses—indicates the investment in biomass needed to keep the animal working in tasks that go beyond its evolved predilections. Wild horses would have eaten grain only sparingly as they grazed the seed heads of grasses; the complete nutrition of amino acids and calories was offered by the meadow alone. Working horses needed additional concentrated energy, and were often kept in climates that did not naturally provide fodder in winter. Stored hay and grains were the stopgap that preserved the domestic horse in the working stable.

The pastoral is powerful because it speaks to our evolved affinities for survivable landscapes, as well as the cultured aesthetic traditions that speak to stewardship, georgic knowledge, and the pleasure of holding land, a measure of wealth. We feel at ease in places with gentle variations, such as mixed woods and grasslands with available fresh water from a running stream or lake. These are the places most likely to provide fuel, shelter, and resources for hunting and gathering. Visual aesthetics of the pastoral show up everywhere from milk cartons to screen savers to street names in subdivisions because they tap into the ecological imagination and suggest health, fecundity, and security. Just as humans seek these locales, Sewell shows the pleasure of the horse in returning to his evolutionary landscape, the pasture.

It is a counterpoint to unhealthy places. The rumpus of London traffic requires an adjustment to modernity: "I had never been used to London, and the noise, the hurry, the crowds of horses, carts, and carriages that I had to make my way through made me feel anxious and harassed; but I soon found that I could perfectly trust my driver, and then I made myself easy and got used to it" (168). Labor as a coal-cart horse is unnatural and killing: "He was sold for coal-carting; and what that is, up and down those steep hills, only horses know. Some of the sights I saw there, where a horse had to come downhill with a heavily loaded two-wheel cart behind him, on which no brake could be placed, make me sad even now to think of" (146). Sewell settles her equines back into emotional and physical health by literally turning them out to pasture. Black Beauty explains: "It was a great treat to us to be turned out into the home paddock or the old orchard; the grass was so cool and soft to our feet, the air so sweet, and the freedom to do as we liked was so pleasant—to gallop, to lie down, and roll over on our backs, or to nibble the sweet grass" (27). The pasture offers freedom from the yokes of labor and the choice of a comfortable place. Agency, the ability to choose one course over others, is available to the horses only at pasture. This affinity of ecological place is best demonstrated when Black Beauty (now "Jack" the London cab horse) spends a Sunday afternoon with his kind master in the country:

> When my harness was taken off I did not know what I should do first—whether to eat the grass, or roll over on my back, or lie down and rest, or have a gallop across the meadow out of sheer spirits at being free; and I did all by turns. Jerry seemed to be quite as happy as I was; he sat down by a bank under a shady tree, and listened to the birds, then he sang himself, and read out of the little brown book he is so fond of, then wandered round the meadow, and down by a little brook, where he picked the flowers and the hawthorn, and tied them up with long sprays of ivy; then he gave me a good feed of the oats which he had brought with him; but the time seemed all too short. . . .
>
> We came home gently, and Jerry's first words were, as we came into the yard, "Well, Polly, I have not lost my Sunday after all, for the birds were singing hymns in every bush, and I joined in the service; and as for Jack, he was like a young colt."
>
> When he handed Dolly the flowers she jumped about for joy. (197)

Everything is present in this pastoral idyll: sun, shade, water, grass, grain, and even superfluous flowers that, by synecdoche, bring the country back to the London stable yard. The scene eases the modern ache caused by the industrialized, machine-driven, crowded life in cities by suggesting that these pleasure spaces coexist with the sphere of business and bustle, just outside the walls of human-centered productivity. The pastoral is the alternative, animated by imagination, ornamented by season, and accessed by animal muscles. The nineteenth-century countryside was refigured by industrial capital and machine transportation into a rural escape for the city-weary, fossil-fueled lives of masters and masses. Lush with sun-knit grass fuel, the pasture is the horse's homecoming.

Black Beauty's last gig—in front of a light gig in the country—is pure joy for a horse that has spent much of his life being mistaken for a machine, specifically a coal-powered steam engine. The source of much of Beauty's suffering comes from the machine-age mentality that a prime mover needs only fuel and maintenance to keep at maximum output. In his experience, men treat horses as machines, and "ninety-nine persons out of a hundred would as soon think of patting the steam engine that drew the train" as the horse that draws their carriage (201). Sewell effectively, if sentimentally, conveys the horse's need for affection and appreciation, as well as fuel and maintenance. From this essential difference, Black Beauty gains perspective on the nature of the locomotive. At first alarmed by its smoky clamor, Black Beauty adapts to living alongside the railroad tracks and keeping calm in his circuits around the station. His initial reaction mimics the testimony of nineteenth-century people alarmed by the coming of the railways to their quiet countryside:

> I shall never forget ... with a rush and a clatter, and a puffing out of smoke—a long black train of something flew by, and was gone almost before I could draw my breath. I turned and galloped to the further side of the meadow as fast as I could go, and there I stood snorting with astonishment and fear.... I thought it very dreadful, but the cows went on eating very quietly, and hardly raised their heads as the black frightful thing came puffing and grinding past.
>
> For the first few days I could not feed in peace; but as I found that this terrible creature never came into the field, or did me any harm, I began to disregard it, and very soon I cared as little about the passing of a train as the cows and sheep did.
>
> Since then I have seen many horses much alarmed and restive at the sight or sound of a steam engine; but thanks to my good master's care, I am as fearless at railway stations as in my own stable. (13–14)

Desensitization aids Black Beauty's calm in this landscape of fuel transition. His quick adaptation captures the stunning capacity of animals to adjust to foreign sensations, and at this fulcrum between biomass and fossil worlds, animals of the field encapsulate the human experience of shifting fuel regimes. Where Thoreau in *Walden* called attention to an indelible change in the Concord, MA landscape with the first sounding of a train horn, Black Beauty is more civic minded. So long as the train did not bring him "any harm," it was easily disregarded.

Experienced in the elision between horse and machine, Beauty wisely seeks to educate ignorant boys, who "think a horse or pony is like a steam-engine or a thrashing-machine, and can go on as long and as fast as they please; ... I just rose up on my hind legs and let him slip off behind—that was all. He mounted me again, and I did the same.... They are not bad boys; they don't wish to be cruel. I like them very well; but you see I had to give them a lesson" (43). Though there is comedy in this figure of the light-hoofed tutor, Beauty is steadily forced to face the reality that humans will continue to abuse his animal essence with mechanical expectations. He makes a caricature of people (class labelled as Cockneys) who employ the "steam-engine style of driving" because they are urbanites accustomed to rail travel, and with little experience of horses. They mistake a seat in a carriage for a rail ticket, and pay no attention to the vagaries of weather, road condition, and topography, expecting "all the same—on, on, on, one must go, at the same pace, with no relief and no consideration" (143). His time as a horse-for-hire is nearly the worst, as "this steam-engine style of driving wears us up faster than any other kind. I would far rather go twenty miles with a good considerate driver than I would go ten with some of these; it would take less out of me" (144). Black Beauty suffers from the ignorance of a bourgeois expectation of service equivalent to money—a ticket that falsely signifies equal labor from inequivalent prime movers. Beauty's era was fraught with confusion over the essence of energy from different engines using different fuel sources.

The ending of the novel is a gentle comedy when Beauty comes full circle back to his original owners, reclaims his name, and lives a pastoral idyll: "Mr. Thoroughgood, for that was the name of my benefactor, gave orders that I should have hay and oats every night and morning, and the run of the meadow during the day.... My troubles are all over, and I am at home; and often before I am quite awake, I fancy I am still in the orchard at Birtwick, standing with my old friends under the apple-trees" (254). His final repose allows him to be simply horsey, with the proper food in the right habitat. Beauty testifies to the anguish of ontological miscategorization—an animal used as a machine—but in the end is spared the death of the common animal in the industrial labor system.

Removed as we are from a close association with animals as transport, I wonder whether our natural affections, rather than being eliminated, have in the last two hundred years been transferred from animals to machines. An animal has a personality, a history, fortunes and misfortunes, and eventually dies. We often assemble the same narrative for our transportation machines, mainly personal cars, but also bikes, trains, and ships. Walt Whitman relocated his theriophily from the animal body—the recurrent subject of *Leaves of Grass*—to the machine body, in the poem "To a Locomotive in Winter." He hails the locomotive as a "fierce-throated beauty" that is "glad, and strong," with "madly-whistled laughter" (ll. 18, 20, 26). My own cars have been named and occasionally patted after a long journey. I have been concerned with their scratches, rust, noises, and their fate at sale—not in the avarice of resale value, but in a human concern akin to the suffering of a living creature. We replace oats and apples with Armor-All and premium gasoline. Perhaps we show mild derangement in our displacement of empathic concerns to machines, one of the symptoms of nature deficit disorder. Our sympathy for inanimate machines identifies a peculiar madness of industrial modernity. Human behavior evolved to be concerned for animals that helped us survive; removed as we are from these relationships, we self-deceive in order to fill an emotional void.

IV Borderline biomass: Peat and whale oil

Peat, or turf, is halfway between biomass and fossil fuel: it is undercooked coal. It comes from bogs, mires that gradually accumulate biomass, mostly sphagnum moss. Through time these wet, acidic moss heaps become carbon sinks. The captured carbon is released again in burning. Peat occupies different fuel categories depending on the monitoring agency. Because it is harvested at rates that far outpace its deposits of about 1 mm per year, it has long been considered a fossil fuel, slow to accumulate and extracted unsustainably. The Intergovernmental Panel on Climate Change (IPCC) defines it as a "slow renewable fuel," a new category between fossil and renewable fuels. Its renewability should not be confused with environmental friendliness: its peat's emission factor—carbon dioxide produced per energy released—is equal to shale oil, greater than all forms of coal, and about twice that of natural gas (Herold 12).

Because of their anaerobic conditions, bogs also preserve archeological relics like human bodies and tree trunks. Digging peat is excavating history, and many of the most informative discoveries about human cultural evolution in the Iron Age, as wells as landscape transformation, have been a side effect of our need for fuel. Peat is culturally resonant in northern Europe, particularly in Ireland where the abundance of bogs and lack of timber make turf digging as essential to Irish domestic energy use as timber to Norway or coal to England.

When it burns in literary hearths, peat almost invariably designates class. It is a fuel of necessity rather than choice, and so it burns in lowly places, and often indicates not only poverty but also cultural identity. In the forgotten 1856 novel *The Daisy Chain* by Charlotte Yonge, the virtuous daughter Ethel means to improve her community by tutoring poor children. Her fastidious brother Norman disapproves of the mission, and the energy source that powers it. Ethel teaches in a small cottage, "a stuffy hole, full of peat-smoke, and with a window that can't open at the best of times," a situation that gives Norman a headache (208). By way of supporting Ethel, their physician father notes that his son's head is more sensitive than his daughter's, and he jokingly opines that "peat-smoke is wholesome" (209). In this little vignette the peat comes to symbolize Ethel's economy and perseverance against the class-based prejudice of Norman, whose moniker represents Anglican domination. He would rather his sister spend her time learning Greek by the clean wood fires at home.

However, sometimes in writings of this period peat is stripped of its cultural stigma and merely demonstrates domestic economy based on available resources. In *Jane Eyre*, the desperately wandering Jane's first impression of St.

John Rivers and his family comes as she spies through their window: "I could see clearly a room with a sanded floor, clean scoured; a dresser of walnut, with pewter plates ranged in rows, reflecting the redness and radiance of a glowing peat-fire" (Bronte 127). At this point in the novel, Jane is desperate for shelter, having fled the roaring wood and coal fires at Mr. Rochester's house. Bronte uses peat to signify the Rivers's prudence, the ecology of their neighborhood, and nothing more. They keep a warm, tidy home despite limited means. A peat fire can illuminate a clean-scoured cottage floor as well as the squalor of a hovel.

The poet Seamus Heaney pays homage to the bog as a locus of deep Irish history steeped in hard labor and the chilly damp dissipated by constant turf fires. Several of his most recognized works visit the bog in order to contemplate the fate of bog men and women, what their preserved deaths reveal about love and revenge, and about the secrets behind the ancient masks of bog-cured faces. The pleasure of hefting the waterlogged turf, of seeing it split by the insistent spade and steam in the sun, is a feeling of union with Irish generations past, Heaney's father most immediately. In his early poem "Digging," Heaney recalls his father digging potatoes and, before him, his grandfather. "My grandfather cut more turf in a day / Than any other man on Toner's bog. / Once I carried him milk in a bottle / Corked sloppily with paper. He straightened up / To drink it, then fell to right away / Nicking and slicing neatly, heaving sods / Over his shoulder, going down and down / For the good turf. Digging" (ll. 17–24). The young man is settling his mind to a writer's life. To access the soil of his ancestry, he fashions his pen into a tool as fit for digging as any other: it needs only a skilled handler. "Between my finger and my thumb / The squat pen rests. / I'll dig with it" (ll. 29–31).

In the same 1966 collection, Heaney's "Bogland" describes the terrain lovingly as "kind, black butter / Melting and opening underfoot, / Missing its last definition / By millions of years. / They'll never dig coal here" (ll. 16–20). America's prairies and England's coal mines have no productive equivalent in Ireland, land of "unfenced country" where the Irish "pioneers keep striking / Inwards and downwards" (ll. 6, 23–24). Dwellers literally sink into their terrain as they excavate its resources. Heaney sculpts a macabre image of the Irish gradually ingested by the land, yet finding their ancestors—human, elk, and fir—alike undigested in the bog, as "every layer they strip / Seems camped on before" (ll. 25–6). They are gathering fuel, and along the way they become archaeologists of their history, paleontologists seeing the primordial past, philosophers spackled in the black butter of Irish energy. Where other countries move across latitudinal frontiers, the Irish dig down into the past, and find a nether horizon in which "the wet centre is bottomless" (l. 28).

In the poem "Kinship," part of a series of bog poems from the 1975 collection *North*, Heaney flows through a series of gentle images that invoke the sensuality of the space: "The bog floor shakes / Water cheeps and lisps / As I walk down / Rushes and Heather" (ll. 9–12). His journey is across the bog, but also down into time, where his imagination greets the primordial "Quagmire, swampland, morass: / The slime kingdoms, / Domains of the cold-blooded, / Of mud pads and dirtied eggs" (ll. 25–8). He has gone past Irish history back to the living ecology of future fossils, considering for a moment each of the forms of living this bog has sustained and swallowed: "Ruminant ground, / Digestion of mollusk / And seed-pod, / Deep pollen-bin" (ll. 31–4). For a fleeting moment he turns to the practical business of harvesting peat:

> I found a turf-spade
> Hidden under bracken,
> Laid flat, and overgrown
> With a green fog.
>
> As I raised it
> The soft lips of the growth
> Muttered and split,
> A tawny rut
>
> Opening at my feet
> Like a shed skin,
> The shaft wettish
> As I sank it upright
>
> And beginning to
> Steam in the sun.

(ll. 49–62)

Gathering turf is at once a practical and a symbolic process. Heaney's mind continually turns literalities into metaphors, as in his description of the turf cart: "The cupid's bow / of the tail-board, / the socketed lips / of the cribs," an act of imagination that transforms the working man into an Irish god (ll. 101–4). No longer a laborer, the poet's god is the archetypal "hearth-feeder" who is delivered "bread and drink" and on the roads is "saluted, / given right-of-way" (ll. 108, 111, 116). Heaney recalls his work as a boy-servant to the turf diggers, a place of privilege and pride that challenges the more celebrated Irish traditions of servitude—and resistance—to Anglo squires. The turf digger's servant is nobler than the duke's butler; together, the boy and digger enact the eco-cultural labor of their land and liberate energy from its heavy bounty.

Unlike globetrotting oil and coal, peat is a regional fuel that evokes a specific locale. Grass is pastoral, often downright idyllic. Grass speaks to the present season, the energy of sunshine falling on the fields. Peat is swampy, at every slice and layer invoking an intimate past history that is still accessible to the present. A coal mine occasionally reveals primordial forms like fern prints, but they exist on evolutionary timescales less accessible to human empathy. Coal's alien forms tend toward abstraction. But peat holds elk horns, intact tree trunks, humans with skin and hair and clothes and identifiable death wounds. Bogs are books turned, leaf by leaf, downward into the folkloric past. This book is often tragic or disturbing. It always speaks of death and loss. But what remains for the present is a clear view of the human condition, squatting on the surface of an engulfing maw, preserving life by burning the half-decomposed bodies of a legible past. Peat's warmth is reluctant and smoky, but in the end it is sufficient. Like the two-thousand-year-old Tollund Man, Heaney and his poems are preserved in turf, "unhappy and at home" (l. 44).

Peat's Irish origins may lend a new perspective on the colonial history with England, since the natural resources found on each island form a poignant contrast. Ireland has bogs that make peat, and looks out on the open Atlantic. England has forests and thick deposits of coal, and its proximity to the European continent has always affected its status as a world power. England's geography, biomass, and fossil fuels push it toward the center of international affairs, with the slave ships made of oak and the revolution made of coal. Ireland's counterparts make it peripheral: a land, like the other colonial lands, that seemed to English ears to cry, "conquerable!" But the wildness of the Irish land, its geology of mucky layers and hidden histories, translates into the Irish resistance to English and, later, British, rule. Looking back on this history from an Anthropocene perspective shows how fuel sourcing can designate a series of cultural departures, by virtue of natural resources becoming cultural identities.

Like peat, which is biomass that behaves like a fossil fuel, whale oil is a tricky fuel to categorize. It is a biomass fuel, but its energy comes not from the strong cellulose fibers forged by plants using solar energy, but from the lipids stored from a whale diet of animals: krill, zooplankton, and sometimes bigger beasts like penguins and squid. In that sense, whale oil is a step higher on the food chain of fuel than the rest of the biomass sources: when we burn wood, the fire eats a vegetarian diet; when we burn whale oil, we are burning the flesh of a carnivore. Predators are always the scarcest beasts in an ecosystem. Animal rights concerns aside, like peat and the forests that provide timber for charcoal, whale oil is a renewable fuel so long as it is harvested at the replenishment rate of whale populations.

Conservationists see in the history of whale hunting a prime exhibit of unsustainability, analogous to the deforestation of Ireland for the wood used in iron forges in the eighteenth century. It is likely a moot point now, because an international moratorium on whale hunting means that very little cetacean energy makes it into human economies anymore. But the history of whaling as an important source of fuel for illumination, and especially the intensive global industry in whale oil from the eighteenth to the mid-twentieth century, requires some consideration in our readings on fuel.

We might question whether whale oil can be considered "fuel" at all. Though whale steaks may have fueled a few human bodies starved of fresh meat on board the *Pequod* in *Moby Dick*, whale oil is not used to feed stock animals, or hearth fires; it does not fuel industrial manufacture, nor does it fill the tank of the moving machine. Whale oil is a specialist fuel that is very good at its appointed tasks: lighting and lubrication. Rather than going in the belly of the nineteenth-century steam engine, as a source of fuel, whale oil eased the friction of its moving parts and kept them lissome and clean. In the days before light bulbs and power stations, whale oil fueled the clearest, brightest, and most fragrant light available.

That pleasing glow came at a price: whale oil was expensive because its acquisition required mighty risk, effort, and sacrifice among its hunters, and its source was increasingly scarce by whaling's heyday in the 1840s. Though it arrived at the docks in thirty-five-gallon barrels, it was sold fractionally in fluid ounces, drams, and scruples. Where coal saved a deforested England from economic ruin in the 1700s, just a bit later, after 1850, kerosene and manufactured gas derived from coal saved the whittled whale populations from the predations of Nantucket. A cornucopian would claim this history as a victory of the free markets, in which rising whale oil prices drove innovation toward another fuel source. Whatever the historical cause of kerosene's arrival, for the sperm whale, kerosene cut down its only predator, the whaleman.

The literature of whale oil is dominated by that colossus, *Moby Dick*. Herman Melville's urtext of whaling instructs in several areas regarding the whale hunt, but oil used as fuel emerges in only a few passages. The state-of-the-art *Pequod* has an onboard tryworks, a stove designed to render fresh whale flesh into liquid fats that could be stored in barrels below decks. When the ship kills and hauls its quarry, the deck is inundated with the meat, guts, and blubber of the beast, and those flows of energy feed the micro-industry that springs up in the wake of the successful hunt. Whalemen cut the floating carcass into long strips of flesh and feed steaks of it into the tryworks pots, where they boil over a fire fed on "fritters"—crackling bits of oily meat—to spare the limited supply of kindling and coal on board.

Always instructive, Melville describes this abundant fuel: "Shriveled blubber, now called scraps or fritters, still contains considerable of its unctuous properties. These fritters feed the flames. Like a plethoric burning martyr, or a self-consuming misanthrope, once ignited, the whale supplies his own fuel and burns by his own body" (374). Stubb, the second mate and captain of the harpoon boat that captured the whale, orders a steak to be cut from the flank. By midnight, he "eats the whale by his own light"—the light of whale oil lanterns (271). Stubb chews to the brutal music of sharks devouring the floating carcass, and complains of the violent din as his own chewing echoes it. Simultaneously, Stubb's belly, the many sharks' bellies, the lanterns, the tryworks fire, and the storage barrels are filled with the fuel from the cetacean, a corporeal counterpart to the "gusher" of petroleum that would thrill the popular imagination in the twentieth century.

The superabundance of pure whale oil is one of the majestic aspects of the whale ship, which glows like a Christmas tree on the open ocean. Melville writes: "But whaleman, as he seeks the food of light, so he lives in light ... See with what entire freedom the whaleman takes his handful of lamps ... to the copper cooler at the try-works, and replenishes them there, as mugs of ale at a vat. He burns, too, the purest of oil, in its unmanufactured, and, therefore, unvitiated state; a fluid unknown to solar, lunar, or astral contrivances ashore" (377). These celestial contrivances are types of lamps that dispel the darkness of Victorian drawing rooms. The land-bound know only the expensive, processed oil found on the shelves of apothecaries, not its original most pellucid form still warm from the rendering fire. Though his work is arduous and filthy, the whaleman's pleasure is the pure light of his raw material, an aesthetic of whaling that makes the practice seem almost divine or supernatural: an infinite supply of light to pierce the inky darkness of the open ocean. Thanks to its indigenous fuel, the *Pequod* is an earth star. But Ishmael scolds the land-locked reader "For God's sake, be economical with your lamps and candles! not a gallon you burn, but at least one drop of man's blood was spilled for it" (194).

These moments of quiet pleasure onboard ship are fleeting, for Captain Ahab's ship is hell-bent on destruction, and an antipode to the ship-as-bright-star comes

just a few pages after the ode to spermaceti. "The rushing Pequod, freighted
with savages, and laden with fire, and burning a corpse, and plunging through
that blackness of darkness, seemed the material counterpart of her monomaniac
commander's soul" (375). The fires are not only the pure whites of lamps; the reds
of revenge bleed into the aesthetic on this ship of fools. When they experience St.
Elmo's fire, Stubb the optimist likens the ship to a great candle saturated with its
resource: "I take that mast-head flame we saw for a sign of good luck; for those
masts are rooted in a hold that is going to be chock a'block with sperm-oil, d'ye
see; and so, all that sperm will work up into the masts, like sap in a tree. Yes, our
three masts will yet be as three spermaceti candles—that's the good promise we
saw" (442–3). Beacon or omen, the glimmering *Pequod* and its captain occupy
a place in Western cultural identity that red-flags obsession and hubris. Besides
Ahab's ambition, the reader swims in the fine details of whaling science, the
nature of its fuel, and the little men who strived anonymously to illuminate a
world before electricity.

Each form of biomass in this chapter has its own nobility and nastiness: the
sweetness of dried hay or the putrid dung it becomes; the butter of peat or its
suffocating sink; pellucid whale oil extracted from an enormous, suffering,
bleeding mammal body. Beyond the aesthetics of the fuel, the literature surveyed
here has uncovered other layers of nature culture. First, the pastoral is a construct
of idealizing urban societies from the Greeks to the early modern courtiers to
the Romantics, and acts as an escape from the social rigors of those spheres.
In more recent, realist literature the pastoral exhibits the physical labor and
finely tuned sense of dwelling that comes from cultivating land that supports
working animals. Both versions of pastoral are romantic, involving love among
shepherds, animals, and landscapes, and the fine-tuning between what the land
and climate provide and what the culture cultivates. This nature culture is the
heart of the texts' emotional arcs.

Next, biomass literature, in common with other fuel sources, signifies social
standing. As peat has been used to degrade its user's standing, and coal is
similarly stigmatized as compared to timber, petroleum, and natural gas; those
who labor to produce animal fodder are members of the lower, laboring classes.
But biomass presents a complication of this well-known stratification in both its
romanticized and realist forms: laborers are personally ennobled by their hard
work, their loyalty to animal and land, their genuine affections and enduring
bonds. Though biomass cultivators may be flagged as the "salt of the earth" by
the literati depicting them, readers can't help but feel some admiration and even
envy of the satisfactions, physical, emotional, and practical, of living within

the sustainable rituals of the steward. This admiration has become all the more fervent with the alienations of the industrial state, but it predates industry as well. Homer and Virgil, Marlowe and Raleigh, Clare and Blake, each honored the biomass laborer. From an Anthropocene perspective, writers like Haushofer and Kunstler had a clear view of the ideals of going retrograde, from fossil fuel and nuclear conditions to the neo-shepherds of a postindustrial condition.

Finally, biomass literature serves as a checkpoint at the border of the fossil-fuel era and the Anthropocene. The nineteenth-century literature that scrutinized this paradigm shift, including DeQuincey, Ruskin, Sewell, and several more writers to come in the next chapter, uses direct comparisons between old ways of transportation—the horse versus the steam engine—to show the skewed ideologies of energy that thrall the fossil-fuel consumer. These works are excellent reminders of the high-energy regime that has become almost invisible by virtue of its ubiquity. None of them means to place hay and fodder at the center of their action and emotion, but each is a thick palimpsest that supports bringing the lower layers into the light. Once enlightened, the contemporary reader can't help but marvel at the energy present that society takes for granted. Turning these elisions into overt and distinct ontologies helps us comprehend the magnitude of our investment in fossil-fuel culture, and marks a few strokes toward imagining an alternative reality. A future biomass regime is potentially both retrograde and progressive.

3

Wood: Forests' Fires

I Burning to survive

Fires convert the chemical energy stored in plants into terrifically useful thermal energy. With the heat from wood, we can cook meat and plants that would otherwise be tough, rancid, or toxic; we can survive cold climates and extend habitable seasons; and we can clear land. With its by-product, light, we can cut through darkness to extend useful time and fend off beasts that would like to eat us. With smoke, we can communicate over distance and difficult terrain, and we can preserve food. Controlling fire is perhaps the most important coordinating element for many features of human evolution: larger brains, smaller jaws, greater omnivory, cooperative hunting, expanded habitat, tool manufacture, and—crucially for a weakling bipedal ape—protection against predators. The tribal fire was the first water cooler, a meeting place that encouraged confabulation, cohesion, laughter, negotiation, memory, and projection. It signified food, safety, light, and warmth. The sayings "home and hearth" and "keep the home fires burning" indicate an ancient intrinsic link between domestic felicity and the fireplace.

In this age of glowing LED screens and phone-based fireplace apps, the affinity between humans and fire is abstracted, perhaps, but this very abstraction underscores humanity's enduring desire for fire. "National Firewood Night," a television program that aired in Norway in 2013, attracted a million viewers—one fifth of the country's population—to watch a live eight-hour broadcast of a burning wood fire. Debates ensued about the right techniques for splitting and stacking wood (bark up or down?), arranging logs in the fire, and cooking wieners and marshmallows. One viewer described the spectacle as "very calming and very exciting at the same time" (Lyall). In Norway, wood fires are necessary for survival, and are symbols of everything they literally provide: safety, warmth, light, company, food, conversation, silence. Once you reach a certain latitude

Figure 3.1 George Inness, *The Lackawanna Valley*, c. 1856. Image courtesy of the National Gallery of Art. Public domain.

away from the equator, certainly at Oslo's sixty degrees North latitude, wood and fire are intimately entwined with life. Take away the need for the actual fire, the emotional desire remains in the nostalgic form of slow-TV viewers clustered around an electronic glow.

In the twenty-first century, people in developed nations spend less time than ever by the hearth, but our fires have only moved and morphed, not extinguished. They burn in our home furnaces, they combust internally in our car engines, they drive turbines in our regional power stations (Wrangham 9). From prehistorical fuel—mostly grass, kindling, and branch wood—humans innovated burning peat, split cord wood, charcoal, coal, whale oil, kerosene, natural gas, and petroleum. In Europe, the source of fuel came to indicate social status: the impoverished burned brushwood gathered in rural areas, and coal in urban ones. The wealthy enjoyed fragrant, bright wood fires fueled by large timber. Oscar Wilde captures the morally barren Sir Henry Wotton, Dorian Gray's chauffeur into degeneracy, by describing his relationship with fuel: "He paid some attention to the management of his collieries in the Midland counties, excusing himself for this taint of industry on the ground that the one advantage of having coal was that it enabled a gentleman to afford the decency of burning wood on his own hearth" (39). Though Wotton's mode of generating wealth is of the new school, by all appearances he lives like an old-realm landed aristocrat, burning fragrant timber even in his London house. The wealthiest landed gentry often had wood brought from their home estates to brighten their seasons in London; burning coal, like for most Londoners of the nineteenth century, was *déclassé*.

The abundance of wood in America was one of the recruiting points for immigrants: wood was abundant to the point of being a nuisance to farming and expansion. Even the poor family off the boat would be sure to enjoy blazing wood fires in the colonies. When the Pilgrims stepped off the *Mayflower*, they escaped the confinement of coal furnaces below decks to find a land "wooded to the brink of the sea." Failing to find the Chesapeake shores veined in gold, the Jamestown colonists sent home cedar logs instead (Freese 259). By the eighteenth century, the American colonies had an endemic population of blacksmiths and colliers whose expertise helped convert the depthless forests into good profits: from forests, to wood, to charcoal, which fired blast furnaces to produce iron. By the time of independence in the 1770s, the United States was producing as much iron as the industrial centers of the world, England and Wales. That stockpile of raw infrastructure came at the price about forty square miles of forest per ton of pig iron (Nye 21). The iron industry, forged from charcoal, enabled westward migration through the vast eastern forests toward the Great Plains, leaving clear-cuts that initiated agriculture.

After millennia of outdoor fires fueled by brush, grass, and fallen logs, humans moved fires into their domestic spaces—caves to castles to condominiums. Architecture from the ancient through modern periods reveals the many ways that indoor fires were burnt. The shapes of rooms, roofs, and hearths changed according to the era's design and the source of fuel. The medieval great room had a fire at its nucleus, and a hole in the roof to allow some modicum of relief from the smoke. These fires could burn anything even remotely combustible, and their nearly continuous ignition required a renewable stack of brushwood.

In the sixteenth century, fireplaces with chimneys began to appear in the walls of buildings, allowing greatly improved ventilation of smoke and the amenities of the hearth. But fireplaces did not burn as well with brushwood fuel from coppiced trees; they demanded larger logs, which raised the demand for good timber (Williamson 231). When coal began to replace scarce timber later in sixteenth-century England, open fireplaces became closed stoves, with narrower chimneys. Coal stoves became ubiquitous in early modern cities in England, which were densely populated, forested with chimneys, and ready for manufacturing industries. Coal stoves had at least three negative effects: more toxic low-level air pollution, the loss of the bright cheery open fire, and a chimney sweep industry that exploited children, who fit in the flues too narrow for adults (Freese 34).

Coal belongs to the fossil-fuel cluster of the next section. This chapter looks into the literature of wood, especially wood fires and their ecological, social, and aesthetic valences. Wood and brush fires still cause deforestation in underdeveloped parts of the world, so this fuel source is an ongoing factor in Anthropocene theory. Though its most obvious resonance is nostalgic in developed countries, wood is also a futuristic fuel: wood pellets are part of sustainable and climate-neutral fuel regimens, especially in northern Europe's timberlands.

Accounts of the age of exploration are replete with images of deep forests in terrains unknown, combining the intrigue of wealth with the exhilaration of danger and mystery. To the colonial explorer, these forests were treasure chests of fuel and raw materials, but they sheltered the uncontrollable elements of nature, hostile natives, predatory animals, poisonous plants, and ruinous diseases. Literature of this early modern period often lingered in foreign forests in order to refresh the wilderness for English readers. In *The Tempest*, Caliban, Shakespeare's cursed island native, bristles against the tyranny of his colonial lord Prospero, but is forced by fear of his torturing Western magic to haul in wood fuel for his fires. Caliban "does make our fire, / Fetch in our wood and serves in offices / That profit us," so he must be tolerated (1.2.456–8). To Prospero, having a fire well supplied and tended is more important than ridding the island of Caliban, a native, angry male attracted to his daughter Miranda. In response to his summons, Caliban's first line is both servile and defiant, a form of anti-colonial back-talk: "There's wood enough within" (1.2.460).

Caliban's intrigue lies not in his ability to make fire, but rather his intimate knowledge of the ecology of the island, a grounded naturalism that rivals Ariel's supernaturalism. Unfortunately for him, the too-trusting Caliban revealed to Prospero the secrets passed from his dead mother Sycorax. Caliban "show'd thee all the qualities o' the isle, / The fresh springs, brine-pits, barren place and fertile: / Cursed be I that did so! All the charms / Of Sycorax, toads, beetles, bats, light on you!" (1.2.488–91). Stripped of exclusive knowledge, Caliban is left in the primitive role of fire maker, and his mind is confined to fantasies of overthrowing his master without the agency to make it happen. He fears too much the "cramps" and "side-stitches" of Prospero's punitive magic (1.2.475–6). Caliban is a sad study of the oppressed colonized man whose technology, the wood fire, is essential for survival but requires hard labor to maintain. As fire maker, Caliban is both abject and necessary. Vin Nardizzi has commented on Caliban's work as fuel-maker: "Caliban keys us into the indispensability of wood as the primary energy source underpinning subsistence and manufacture in the preindustrial era. Moreover, the response encodes a fantasy of plenty articulated during a time of shortage in England" (313).

Robinson Crusoe's domain is another fantasy colonial forest, arrived at by shipwreck. The famed survivalist follows his instinct from the first: "Having now fixed my habitation, I found it absolutely necessary to provide a place to make a fire in, and fuel to burn" (50–1). Crusoe finds shelter, food, security, and the companionship of goats, but the visible aspects of the wood fire become a danger for the anxious castaway. After spotting a footprint in the sand, he is obsessed with the fear that the smoke will attract cannibals to his "castle," and he eventually retreats to his inland cave "grotto" to burn raw wood: "I could not live there without baking my bread, cooking my meat, &c.; so I contrived to burn some wood here, as I had seen done in England, under turf, till it became chark or dry coal: and then putting the fire out, I preserved the coal to carry home, and perform the other services for which fire was wanting, without danger of smoke" (138). For a character dwelling in a land of infinite fuel, Crusoe kindles surprisingly few wood fires. His use of charcoal and gunpowder will receive attention later in this chapter.

Jack London's "To Build a Fire" develops the elemental necessity of fuel in an age of realism on a new frontier: Alaska. This story is about bare-bones survival, the danger of exposure, and the narrow margins in the physics of burning. A desperate man is far from camp on a subzero day with wet feet, so he must stop to warm up and dry out. But incubating a fire in such terrain is a ticklish process:

> On top, tangled in the underbrush about the trunks of several small spruce trees, was a high-water deposit of dry fire-wood—sticks and twigs, principally, but also larger portions of seasoned branches and fine, dry, last-year's grasses. He threw down several large pieces on top of the snow. This served for a foundation and prevented the young flame from drowning itself in the snow it otherwise would melt. The flame he got by touching a match to a small shred of birch-bark that he took from his pocket. This burned even more readily than paper. Placing it on the foundation, he fed the young flame with wisps of dry grass and with the tiniest dry twigs. (15)

Though the scene is primordial, the traveler's expertise in combustion is advanced. It exhibits the delicate dance among various forms of carbon matter, air, water, and extremes in temperature—from the ambient seventy-five below zero to the incendiary Fahrenheit 451:

> He worked slowly and carefully, keenly aware of his danger.... . When it is seventy-five below zero, a man must not fail in his first attempt to build a fire— that is, if his feet are wet. If his feet are dry, and he fails, he can run along the trail for half a mile and restore his circulation. But the circulation of wet and freezing feet cannot be restored by running when it is seventy-five below. No matter how fast he runs, the wet feet will freeze the harder. (15)

All the complexity of human life crystalizes in this life-and-death moment. The traveler's noisy brain brings into further relief the stark clarity of his bifurcated path into the future: fire and life, or ice and death. Seen from a materialist perspective, the reluctant fire possesses more agency than its unnamed builder: he is at the mercy of uncooperative matter. On the brink of failure, the traveler desperately makes one last effort to kindle the fuel:

> He cherished the flame carefully and awkwardly. It meant life, and it must not perish. The withdrawal of blood from the surface of his body now made him begin to shiver, and he grew more awkward. A large piece of green moss fell squarely on the little fire. He tried to poke it out with his fingers, but his shivering frame made him poke too far, and he disrupted the nucleus of the little fire, the burning grasses and tiny twigs separating and scattering. He tried to poke them together again, but in spite of the tenseness of the effort, his shivering got away with him, and the twigs were hopelessly scattered. Each twig gushed a puff of smoke and went out. The fire-provider had failed. (19)

The fire fizzles out, leading to an infinite cold night for the condemned traveler. His dog survives, though, linking this micro-drama with the larger world. The dog is in effect the focalizer who bears witness to the failed fire provider, becoming the realist writer's alternative archetype to the fortunate frontiersman. London's story exhibits how often elements and chance win out over human planning, experience, and effort. The cosmos is indifferent, and any romantic notions about the wilderness are extinguished by London in his tale of failed fuel, or cold wood. Though the story depicts a single man's failure, his travail exemplifies the common struggle for survival over tens of thousands of years, as humans gradually gained footing in hostile climates. As we learned to manipulate the tool of fire, we radiated south and (especially) north to the brow of the globe, lengthened habitable seasons, extended useful light past sunset, and expanded the variety of foods we could eat. The reader, perhaps snuggled in a furnace-warmed home with a warm beverage, is by imagination stripped down and back to biomass fuel, an essential ingredient of evolution. Kindling evokes the thin margins of our ancestors, easily scattered and snuffed out.

II The hospitable blaze

The *ubi sunt* motif usually laments the passing of a beloved person. But an ecological version of the trope appears in the literature of wood during the long age of deforestation. England was covered with forests in 4000 BC, but after the arrival of Neolithic man, these forests were steadily converted to agricultural land. By 500 BC only half of England was forested (Freese 259). Humans usually value a resource only when it is scarce, so the literature of wood in England is particularly rich in its period of rising endangerment: the seventeenth to nineteenth centuries. Wood fires and the comfort of the open hearth appear often in scenes of hospitality, since they measure the wherewithal and generosity of the host. In his great house poem *To Penshurst*, dating from 1616, Ben Jonson celebrates the fires at the heart of hospitality:

> Thy tables hoard not up for the next day;
> Nor, when I take my lodging, need I pray
> For fire, or lights, or livery; all is there,
> As if thou then wert mine, or I reigned here:
> There's nothing I can wish, for which I stay.
> That found King James when, hunting late this way
> With his brave son, the prince, they saw thy fires
> Shine bright on every hearth, as the desires
> Of thy Penates had been set on flame
> To entertain them; or the country came
> With all their zeal to warm their welcome here.

> (ll. 71–81)

Embracing the aristocratic bounty, Penshurst's fire and lights provide the bodily comforts of warmth and perspective, and the livery means clean clothes for the travelling guest or, by synecdoche, servants dressed in livery who wait upon his demands. The estate is not only the building, but also its private ecological amenities: strolling grounds and game-sheltering, fuel-filled forests. The hospitality is so complete that the guest himself feels as a sovereign over Penshurst. The estate served as a refuge to James I and his doomed son, Charles I, who brought their kill to its hearths and enjoyed the service of the staff, figured as

Penates, Roman mythical gods of the household, whose Latin root, *penus*, means provisions. The literal wood fire that warms Jonson doubles as a metaphor for English hospitality, tradition, and patriotism. Since the ode is addressed to the estate, the actual bodies at work to keep the fires lit and the provisions stocked are obscured and subsumed into the aura of Penshurst. Penury haunts the dark margins of existence, but Penshurst expansively glows with health and safety. Fueled by its ample timber resources, it is a refuge in the struggle for survival, for king and poet minstrel alike.

Fire is too evocative to stay in the fireplace for long; it spreads figuratively into memory, desire, and rage. Derek Walcott comments on the legacy of English great houses and the colonial spoils that financed them in his 1962 poem "Ruins of a Great House." He bitterly speculates on the imperial slave economy that supported the grandeur of the house: "The world's green age then was a rotting lime / Whose stench became the charnel galleon's text" (ll. 35–6). Indignation carries his thoughts from corrupt wood to the coffin of the slave ship to the fire of reflective rage:

> As dead ash is lifted in a wind,
> That fans the blackening ember of the mind,
> My eyes burned from the ashen prose of Donne.

> Ablaze with rage, I thought
> Some slave is rotting in this manorial lake,
> And still the coal of my compassion fought
> That Albion too, was once
> A colony like ours.

(ll. 38–45)

His meditation shifts to Donne's Meditation XVII, "No man is an island entirely of itself; every man / is a piece of the continent," and he seems to forgive the ruined Great House for its legacy because it, too, was a victim of colonization and "bitter faction" in English history (l. 41). From wood fuel to burning anger to the quelling balm of Donne's words of unity, Walcott's terse, rough lyric concludes with the fragment quote: "as well as if a manor of thy friend's ..." (l. 44). Donne's meditation ends: as / well as if a Manor of thy friends or of thine / owne were; any mans death diminishes me, / because I am involved in Mankinde; / And therefore never send to know for whom / the bell tolls; It tolls for thee." Though the sight of the Great House fuels the fire of resentment in the colonized man, what remains in the end are the ashes of empathy and compassion.

Even Dracula, literature's denizen of the cold and dark, knows the emotional importance of fire to mortal humans. His guest Jonathan Harker is frigid with fear as he arrives at the vampire's castle. The coachman who bore him to safety through dark Transylvania has evanesced, and in his place appears Harker's uncannily similar host. Like the coachman, his handshake is "cold as ice—more like the hand of a dead than a living man" (22). No other person appears to greet him. However, when the count leads Harker to his chambers, he "rejoiced to see within a well-lit room in which a table was spread for supper, and on whose mighty hearth a great fire of logs, freshly replenished, flamed and flared" (23). Later, when he is given a castle tour, Harker similarly finds "a great bedroom well lighted and warmed with another log fire—also added to but lately, for the top logs were fresh—which sent a hollow roar up the wide chimney" (23). Surely these well-attended fires, and his well-roasted chicken suppers, imply the presence of a mortal servant?

Yet there is none. For a skittish Englishman spooked by the gothic darkness of the castle and harrowed by the cold, wolf-infested woods on all sides surrounding it, these hearth fires signify a great deal more than chemical combustion. They represent safety, wealth, nourishment, and company. Remarkably, the simple gesture of the roaring wood fire allows Harker to perceive the chilly Count Dracula as a host who provides a "courteous welcome," filled with "light and warmth" (23). The Transylvanian (literally "cross-forest") wood transforms from a wolfish menace to a natural resource of a quality long gone in England. Rather than making the castle hollow and gothic, it makes it convivial and romantic. Harker is able at first to brush aside the spookier aspects of firelight— its flickering shadows that distort facial features and spatial dimensions. Wood fires forestall terror for precisely one night, this first one of Harker's residence. After that no more fires are mentioned, and the only blaze comes in the form of a "demonic fury" in the count's eyes when Harker cuts himself shaving (33).

This juxtaposition of warmth with chill, the domestic versus the wild, is a leitmotif in all kinds of spooky stories. Literary works built upon suspense, terror, and horror must ameliorate these stressors with relaxants, and a crackling wood fire and clear daylight are the two best anodynes. In Washington Irving's "The Legend of Sleepy Hollow," the fluttery hero Ichabod Crane is a schoolmaster who moonlights as a biomass handyman. He "assisted the farmers occasionally in the lighter labors of their farms, helped to make hay, mended the fences, took the horses to water, drove the cows from pasture, and cut wood for the winter fire" (297). These small services secure his spot by the fire with tale-spinning Dutch wives, a band of friendly witches who inhale the aromas of roasting apples and exhale local ghost stories. Ichabod would exchange his "anecdotes of witchcraft ... direful omens and portentous sights" for their "marvelous tales of ghosts and goblins, and haunted fields, and haunted brooks, and haunted bridges, and haunted houses, and particularly of the headless horseman, or Galloping Hessian of the Hollow" (299). These fireside chats capture an early American gothic where even the globe's inherent spin takes on a supernatural weirdness that blends physics with folklore.

The fire sitters work each other into a frenzy of apprehension, diffused at the moment by the company and the companionable fire, but stored for later haunting in the imagination of the superstitious Ichabod. In early nineteenth-century New York State, nighttime travel was a lonely road, and the hero's pleasurable sojourn becomes a harrowing travail of his imagination's weaving: "But if there was a pleasure in all this, while snugly cuddling in the chimney corner of a chamber that was all of a ruddy glow from the crackling wood fire, and where, of course, no spectre dared to show its face, it was dearly purchased by the terrors of his subsequent walk homeward. What fearful shapes and shadows beset his path, amidst the dim and ghastly glare of a snowy night!" (299). Ichabod prevails over the mind spirits and reaches home this time, but the episode sets up the story's famous climax in which he is bested by a pumpkin. Irving's amusing satire of horror adopts the conventions of the gothic from eighteenth-century British literature, but he tailors them for an American landscape in which the wood fire is the nucleus of comfort and survival. The woods are spooky, dark, and deep. The settler's essential work is to harvest from a woodsy wilderness a neat split stack at the cabin door. The woodpile holds the promise that civilization can carve enough energy out of the indifferent, cold, dark, material universe.

Because it required so much strenuous work and symbolized settlement, Robert Frost's poem "The Wood-Pile" is spooky because its very presence indicates an absence. A traveler crosses that archetypal wilderness, the swamp, in a frozen winter wasteland. His footing on thin ice is unsure and the trees are "Too much alike to mark or name a place by / So as to say for certain I was here / Or somewhere else: I was just far from home" (ll. 7–9). Still, this traveler is looking for an experience; his imagination peoples the wild. He makes a companion of a skittish bird who alights on an incongruous "pile of wood":

It was a cord of maple, cut and split
And piled—and measured, four by four by eight.
And not another like it could I see.
No runner tracks in this year's snow looped near it.
And it was older sure than this year's cutting,

Or even last year's or the year's before.
The wood was gray and the bark warping off it
And the pile somewhat sunken. Clematis
Had wound strings round and round it like a bundle.

(ll. 23–31)

With the woodpile serving as a text, the traveler reads the ghostly verbs of the vanished settler, who has cut, split, and stacked. The traveler perceives the craftsman settler's experience in creating the clean geometry of the woodpile, four by four by eight, which is reinforced by Frost's steady pentameter. Nature has, in turn, shown a paradoxical care for neglect in the intervening years. The pile is clean of tracks and tied in a neat bundle of clematis, a biomass embrace. Unlike human care, which tends to preserve form, Nature's care is a conversion process, and the decay toward a new form shows in its vanishing geometry and warped bark.

Slowly, the pile of civilization is digested by natural entropy. The energy expended to cut, split, and pile has not served its laborer, but it has assisted nature in the process of turning deadwood into recyclable nutrients by increasing surface area. The fire is starved, but the termites and mushrooms feast. The traveler attempts to know the vanished settler through this strange vision of neglected work; he must be a man "who lived in turning to fresh tasks / Could so forget his handiwork on which / He spent himself, the labor of his ax" (ll. 35–7). Like the traveler in an antique land observing the ruins of Ozymandias, Frost pities the woodpile for its irrelevance, and finds irony in a crumbling ruin that displays consummate skill. Instead of serving its primary purpose in a fireplace helping its maker survive, the pile works at the simmering pace of compost "To warm the frozen swamp as best it could / With the slow smokeless burning of decay" (ll. 39–40). The settler's neglect shifts to nature's loving attention. Like Wallace Stevens's jar in Tennessee, the woodpile defines the wilderness surrounding it; unlike the inorganic glass, the woodpile is itself becoming wild. Within years, thanks to the tireless work of termites and twining vines, there will be no trace of its maker's hard labor.

III In depth: Jane Austen

Considering the crucial role that the wood fire has played in human social evolution, it is not surprising that eagle-eyed Jane Austen would see the fireplace as an ideal location to test character and indicate wealth, that is, access to resources. That wood was rapidly being replaced by coal in domestic uses during her lifetime only enriches the piquancy of her observations. Her novels elaborate on how this seeming background element in a regency room is actually an important leitmotif that Austen deploys with characteristic wit and ruthlessness.

In Europe, since medieval times, and especially during the Industrial Revolution, the source of domestic fuel came to indicate social status. Rural peasants usually burned brushwood gathered in the commons, and the poor burned cheap coal in cities. The wealthy enjoyed fragrant, bright wood fires fueled by large timber at their country estates and imported wood to their London townhouses for use in the kitchen, since coal left a nasty taint on the taste of food. Due to chronic deforestation and the great increases in mineral mining in the Midlands, coal began to replace wood in English fireplaces surprisingly early. In London, by the mid-seventeenth century, coal was commonly the domestic fuel across the classes. One hundred years later, coal burning was common in city and country alike, though its use was influenced by access to the sea or load-bearing canals. In the English interior, wood and brush continued to be valuable resources for domestic heating well into the nineteenth century, and the poor would scavenge the countryside for free fuel with increasing desperation as common lands were enclosed. Children were often assigned the task of gathering brushwood from public lands: F. M. Eden in his 1797 book *The State of the Poor* laments the "young marauders" who ravaged hedgerows and copses to supply their families with fuel for warmth and cooking (Adkins and Adkins 95). Literary urchins Samuel Coleridge and Dorothy Wordsworth gathered brush for their fireplace on the Quantock hills near Nether Stowey in 1798, which Wordsworth notes in her journal as a quotidian domestic chore for the country resident (Adkins and Adkins 96).

For Jane Austen, writing in this time of timber scarcity and accelerating coal consumption, the wood fire is a kind of pigment that helps her illustrate and color different human types and the resources available to her socially stratified characters. Since most of Austen's scenes take place in the country among the upper classes, it is not unreasonable to assume that the fuel is wood. Notably she only specifies the fuel source, coal, when it is a class-based observation of Fanny Price's impoverished family in *Mansfield Park*. Lacking her clear designation, we must assume from location and consumer that most of Austen's domestic hearths were wood-fired.

Austen's characters are variably described as domestic, generous, gregarious, silly, naïve, rich, selfish, stingy, lazy, hypochondriacal, scheming, or sleepy according to their relationship to the wood fire. *Pride and Prejudice*'s Mr. Bingley, for example, demonstrates his generosity, attentiveness, and growing love for Jane Bennet in the way he tends the Netherfield fireplace: "He was full of joy and attention. The first half-hour was spent in piling up the fire, lest she should suffer from the change of room; and she removed at his desire to the other side of the fireplace, that she might be further from the door. He then sat down by her, and talked scarcely to anyone else" (40). Through Bingley's cavalier use of Netherfield's resources, we detect that his natural exuberance can tend toward the silly, so smitten is he by his invalid guest. But Austen's teasing account of his behavior is laudatory rather than mocking. Bingley's fires burn clean and true to his good heart.

The fireplace is an orientation marker in Austen's rooms, just as important as the windows and doors. Her characters approach it, pace by it, gather around it, draw away from it, gaze into it, lean and pose on its mantel. It is the place to get warm after a damp journey, to share confidences, to draw information from a knowledgeable companion, to admire a collection of china baubles, to burn letters, to toast successes. Sometimes Austen's use of fire is not so much a delineation of character as it is a record of the physical strains of travel and exposure that were realities in the early nineteenth century. When the Dashwood sisters travel by carriage to London in *Sense and Sensibility*, we fully understand their physical relief when "they reached town by three o'clock the third day, glad to be released, after such a journey, from the confinement of a carriage, and ready to enjoy all the luxury of a good fire" (182). Three days of jostle and Mrs. Jenning's jabber merit the refreshment gained from a long bask by the "luxury" fire, presumably made of wood despite its urban setting.

More artfully, Austen plays out a dynamic between Emma's father Mr. Woodhouse and the love-struck Mr. Elton, by way of fire. Annoyed at having a social engagement on a cold Christmas evening, the feeble father strikes out against his hosts: "'A man,' said he, 'must have a very good opinion of himself when he asks people to leave their own fireside, and encounter such a day as this, for the sake of coming to see him. He must think himself a most agreeable fellow; I could not do such a thing. It is the greatest absurdity—actually snowing at this moment! The folly of not allowing people to be comfortable at home, and the folly of people's not staying comfortably at home when they can!'" (107). Division from the fire represents a threat to the neophobic old man, and he comically turns his own fears of exposure into an accusation of egotism in the modest Mr. Weston. In response, the smitten Mr. Elton reassures Emma's father by employing the apposite domestic image: "We are sure of excellent fires," continued he, "and every thing in the greatest comfort. Charming people, Mr. and Mrs. Weston" (110). His ardour rolls on, until it gets stuck in the snow with an unwilling Emma.

Outdoors, the aesthetic perception of smoke from chimneys indicates the mood and disposition of the perceiver and his or her subjective impression of the landscape. Emma begins to view Robert Martin with greater favor when she evaluates the utility and beauty of his Abbey Mill Farm: "The considerable slope, at nearly the foot of which the Abbey stood, gradually acquired a steeper form beyond its grounds, and at half a mile distant was a bank of considerable grandeur, well clothed with wood ... It might be safely viewed with all its appendages of prosperity and beauty, its rich pastures, spreading flocks, orchard in blossom, and light column of smoke ascending" (338). This pastoral idyll is a working-class version of Elizabeth Bennet's arrival at Pemberley in *Pride and Prejudice*, which I will discuss below. Martin's farm is not just a site of labor; it is filled out by time-honored buildings, picturesque land contours and, importantly, natural resources. The bank "well clothed with wood" implies ecological health and sustainability, and the management of this resource feeds directly into the "light column of smoke" rising from the farmhouse chimney.

This happy aesthetic medium between beauty and function is also a reassuring social medium between wealthy landowners (the stalwart Knightly owns all the land in sight) and the honest, worthy worker (Robert Martin) who keeps the land groomed and profitable. Wealth *and* work cooperate in this scene: the standing woods and Knightly; the column of smoke and Martin. The smoke itself has an aesthetic. It echoes a feature of a Grecian temple: a "light column," naturalized. It is singular in the landscape, giving texture to the clear-aired day rather than polluting it. It is light in color (unlike coal smoke) and light in density (not urban industrial). Emma's misperceptions are famous, but in this case we view through her an accurate ideal of the pastoral scene, sustained by the right stewards. These rare long scenes of landscape description in Austen usually indicate the swing of a character's mind toward new perspectives. Here, the farm's woods and its domestic fires help Emma's imagination to settle her friend Harriet there as Robert's wife.

Austen presents a counterpoint to Emma's pleasant musings in *Persuasion*, where the tormented heroine Anne identifies her discontent with her place of forced habitation, Bath. Anne "persisted in a very determined, though very silent disinclination for Bath; caught the first dim view of the extensive buildings, smoking in rain, without any wish of seeing them better; felt their progress through the streets to be, however disagreeable, yet too rapid; for who would be glad to see her when she arrived? And looked back, with fond regret, to the bustles of Uppercross and the seclusion of Kellynch" (110). Anne suffers from the persistent melancholia of "fond regret," or toxic nostalgia. Her mind is continually trapped in the past rather than engaging the present, as evidenced by her disinclination to see Bath better. The dour scene is made dim with the "extensive buildings, smoking in rain," which imply both a repellant density of habitation and an unhealthy battle between indoor fires and outside wet chill. The smog, smoke, and damp induce a feeling of miasma and claustrophobia, psychosomatic though it may be. The buildings themselves are smoking, not the chimneys, implying a ground-level pollution that sullies lungs. Anne is unhappy during the journey, and she anticipates further misery when she arrives. Where Emma enjoyed wide perspectives, landscape variation, fine weather, and a complete view of the workings of the system (both ecological and interpersonal), Anne endures all the opposite: obscure, jumbled views passing too rapidly, monotonous urban construction, British weather further dirtied by smoke, and a keenly felt ignorance of locale. Austen chooses the domestic wood fire as a distinguishing marker in these subjective portraits of her contrasting heroines.

With Catherine Morland, the young innocent who wanders wide-eyed down the halls of Northanger Abbey, the pleasure and sense of safety afforded by a live fireplace is a relief from her gothic imagination. The spooky ambience of the abbey haunts her every perception, and she is assailed by a "tempest" upon arrival, which "brought to her recollection a countless variety of dreadful situations and horrid scenes, which such buildings had witnessed, and such storms ushered in; and most heartily did she rejoice in the happier circumstances attending her entrance within walls so solemn!" (121–2). Morland's internal monologue blends sensational fiction with her sights of the abbey, and she buttresses her will: "*She* had nothing to dread from midnight assassins or drunken gallants.... In a house so furnished, and so guarded, she could have nothing to explore or to suffer ... Thus wisely fortifying her mind, as she proceeded upstairs, she was enabled ... to enter her room with a tolerably stout heart; and her spirits were immediately assisted by the cheerful blaze of a wood fire" (122). The happy fire dissipates her worst fears derived from gothic novels: "How much better to find a fire ready lit, than to have to wait shivering in the cold till all the family are in bed, as so many poor girls have been obliged to do, and then to have a faithful old servant frightening one by coming in with a faggot!" (122). We might smirk at the terror of a girl who cannot make her own fire but needs a servant to frighten her first before she can be comforted. The preexisting fire burns away her superstitions about shadowy abbeys and shows that the Tilneys are gracious hosts who anticipate the desires of their guest. If Catherine has a fire in her room, she is respected; at least, she is worth the spent resources.

Fanny, the poor urchin of *Mansfield Park*, receives no such respect from her guardians. As much as the servant's dress she is given to distinguish her as inferior to her cousins, the distinction of rank is signified by fire, or the lack thereof: "Mrs. Norris, having stipulated for there never being a fire in it on Fanny's account, was tolerably resigned to her having the use of [the room] nobody else wanted, though the terms in which she sometimes spoke of the indulgence seemed to imply that it was the best room in the house" (120). Mrs. Norris, monster of this gothic house, controls the domestic economy by doling out unequal favors. Fanny takes advantage of Norris's intolerance by growing stoic and modest, a contrast to her spoiled female cousins. Her absentee uncle Lord Bertram discovers Mrs. Norris's petty tyranny when he returns from his slaveholding estate in the West Indies:

> It was indeed Sir Thomas who opened the door and asked if she were there, and if he might come in. The terror of his former occasional visits to that room seemed all renewed, and she felt as if he were going to examine her again in French and English.
>
> She was all attention, however, in placing a chair for him, and trying to appear honoured; and, in her agitation, had quite overlooked the deficiencies of her apartment, till he, stopping short as he entered, said, with much surprise, "Why have you no fire to-day?"
>
> There was snow on the ground, and she was sitting in a shawl. She hesitated.
>
> "I am not cold, sir: I never sit here long at this time of year."
>
> "But you have a fire in general?"
>
> "No, sir."
>
> "How comes this about? Here must be some mistake. I understood that you had the use of this room by way of making you perfectly comfortable. In your bedchamber I know you *cannot* have a fire. Here is some great misapprehension which must be rectified. It is highly unfit for you to sit, be it only half an hour a day, without a fire. You are not strong. You are chilly. Your aunt cannot be aware of this." (248)

By way of the vacant fireplace, the injustice is revealed to Sir Thomas in a way that Fanny could never have told him, in keeping with her uncomplaining character. Fire fosters survival in basic subsistence, and even in these more velvety conditions it supports health and connotes welcome. By depriving Fanny a fire, Mrs. Norris symbolically effaces her existence in the house: her chimney vents no smoke. Sir Thomas rectifies the rudeness in the same chapter:

> She was struck, quite struck, when, on returning from her walk and going into the East room again, the first thing which caught her eye was a fire lighted and

burning. A fire! it seemed too much; just at that time to be giving her such an indulgence was exciting even painful gratitude. She wondered that Sir Thomas could have leisure to think of such a trifle again; but she soon found, from the voluntary information of the housemaid, who came in to attend it, that so it was to be every day. Sir Thomas had given orders for it. (255)

Fanny's position is strengthened, and Mrs. Norris is chastened, by the luxury of the colonial family burning wood fires in the estate's many fireplaces. Even the attending housemaid seems to enjoy this turn of Fanny's fortunes, though it makes more work for her.

The fire is one element of the distinction of wealth and breeding that Fanny feels acutely when she returns home to her poor nuclear family in Portsmouth, where there is more fire drama:

"Dear me!" continued [Fanny's] anxious mother, "what a sad fire we have got, and I dare say you are both starved with cold. Draw your chair nearer, my dear. I cannot think what Rebecca has been about. I am sure I told her to bring some coals half an hour ago. Susan, you should have taken care of the fire."

"I was upstairs, mama, moving my things," said Susan, in a fearless, self-defending tone, which startled Fanny. "You know you had but just settled that my sister Fanny and I should have the other room; and I could not get Rebecca to give me any help." (301)

The housemaid, Rebecca, is inattentive; the sister, Susan, is saucy, and they all sit in the chilly misery of an inadequate fire. Worse, it is a coal fire: possibly an indication of their closeness to the ocean where sea coals were gathered, but certainly a sign of their impoverishment. The Price daughters feud over their chores in a display of impudence that shocks the aristocratically trained Fanny. This passage is unique among Austen's works in naming the fuel source, a piece of information that carries geography and class alongside the sadder aesthetics of smelly and polluting coal fires. Its very existence shows by contrast the luxuries taken for granted in the rest of Austen's canon, where Fanny has observed a norm of good fires made of rare, expensive wood that has been chopped, split, and prepared by servants.

In *Pride and Prejudice*, the fireplace illuminates a contrasting variety of wooing young men: Bingley, Collins, and Darcy. During his doomed first proposal to Elizabeth, Darcy positions himself in a lord-of-the-manor pose, "leaning against the mantel-piece with his eyes fixed on her face" (146). Bingley, as seen above, reveals the size of his heart by the high pile of wood in his fireplace at Netherfield. He is eventually rewarded for his affability by a more intimate tête-à-tête than

Darcy could at first achieve: Elizabeth "perceived her sister and Bingley standing together over the hearth, as if engaged in earnest conversation; and had this led to no suspicion, the faces of both, as they hastily turned round and moved away from each other, would have told it all" (264). In her various strategies to marry off her too many daughters, Mrs. Bennet quite literally stokes the fire in order to manipulate the passion of Mr. Collins, who "had only to change from Jane to Elizabeth—and it was soon done—done while Mrs. Bennet was stirring the fire. Elizabeth, equally next to Jane in birth and beauty, succeeded her of course" (53). Austen subtly inserts the fire metaphor to brilliant effect: Bingley's claim upon Jane's affection must not snuff out Collins's interest in the Bennet daughters; the mother stirs away those ashes and brings fresh oxygen to the ample fuel in the hearth. Mrs. Bennet has five logs to burn, and they are piled in all at once, choking each other off. The anxious mother sets Elizabeth squarely in Collins's flame, but alas it is too feeble to ignite our weighty heroine.

An inferno on Pemberley's scale is needed to set her ablaze. Elizabeth observes that "to be mistress at Pemberley might be something!" only after she views the estate from the Gardiners' slow-rolling carriage (185). The extended description begins, "The park was very large, and contained a great variety of ground. They entered it in one of its lowest points, and drove for some time through a beautiful wood, stretching over a wide extent" (185). The sylvan state of Pemberley not only emphasizes Darcy's good taste in landscape, but perhaps more importantly also the security of his wealth. Darcy's standing reserve of timber is harvested sustainably to keep his fireplaces well-fueled, and the forest remains intact because of his prudent management. Elizabeth's playful avarice in regarding Pemberley as "something" finds a central symbol in fuel wood, which plays a role both outdoors in the park and indoors in the impressive fireplaces— timber exhibits Pemberley as an exceptional property.

Mr. Collins is less interested in the fire itself than in the wealth demonstrated by its framing. Always close to offending everyone, Collins tells Elizabeth's Aunt Phillips that her main drawing room resembles the "small summer breakfast parlour at Rosings; a comparison that did not at first convey much gratification," and goes on to enumerate why the likeness is complimentary: "When she had listened to the description of only one of Lady Catherine's drawing-rooms, and found that the chimney-piece alone had cost eight hundred pounds, she felt all the force of the compliment, and would hardly have resented a comparison with the housekeeper's room" (57). Collins's cupidity is infectious among the trivial-minded characters, and Rosings's chimneypiece illustrates Lady Catherine's force of character and her riches, which dazzle the clownish crowd: Collins, Mary,

Aunt Phillips, and Sir Lucas. Like Darcy's, Lady Catherine's blazes are built of the finest dried hardwood harvested from her private wooded estate in Kent. It is surprising that the lady fanfaron never proclaims this advantage to her guests. Since the wood represents indigenous resources rather than specific monetary value (like her glazed windows and chimneypiece), the luxury of its abundance may escape her notice.

In each of these cases, except the poor Prices who burned coal in Portsmouth, the very fact that Austen's gentry families burn wood is an indication of their wealth. By the eighteenth century, the English groves that sheltered Robin Hood had been significantly diminished by centuries of deforestation. The forests of England had long supplied the navy with ships (Nelson's HMS Victory required six thousand oaks) and the iron furnaces with wood for charcoal, had fed domestic hearths, and provided the raw materials for the construction of buildings, bridges, farm equipment, and vehicles. To a certain extent, coppice groves were managed sustainably, supplying brushwood for fuel, but that was not the kind of large split lumber desired for the aristocrat's stately blaze. Peering through this specific lens of fireside hospitality, we find a sociohistorical trope that Austen uses as a rasp to chisel her character's features.

Now most often superfluous because of the furnace, the fireplace continues to represent wealth in today's homes. We use electricity or natural gas to heat our houses, so the fireplace now occupies the aesthetic seasonal role of coziness. As discussed earlier, the satisfaction of a fire has evolutionary roots. We still go retro with fireplaces because society has evolved away from open fires faster than biological evolution can adjust our innate preference for their light and warmth. Burning wood for pleasure is not exactly villainous in the Anthropocene, but it does have environmental costs. It drains out the carbon sink of the timber itself and contributes local air pollution in the form of soot, which increases the ratio of heat-absorbing dark material in a cold land of white, reflective snow and ice. Still, it is a less dramatic energy indulgence than turning the key in an ignition. Any guilt we may feel burning wood is no longer associated with a limited supply that implies deforestation, as it would have a few centuries ago. Now, the guilt is caused by the larger process of climate change in the Anthropocene, where every emission contributes to a global process within our collective agency but beyond our individual control. The pure fireside pleasure of Austen's era has become an ambivalent negotiation between coziness and climate burden. The inscrutable line of causation between individual actions and global climate patterns captures the essence of the climate hyperobject, and the psycho-emotional condition of living in this century of ripe consequences.

4

Charcoal: Firing Early Industry

I The collier

Charcoal is the biomass fuel closest to a fossil fuel. It looks and behaves quite like its fossil counterpart, coal: black, brittle, smeary, dusty, and energy dense. Charcoal is a slightly more concentrated fuel than bituminous coal, containing about a third more millijoules per kilogram, and it has twice the energy density of well-dried wood (Smil 12). It is processed by slow anaerobic burning of wood, or pyrolysis: literally a "separating fire" that eliminates water as it preserves the vasculature of the wood (its veins) so that a light, porous fuel remains. This distillation of wood energy was the first true chemical process developed by humans in the Bronze Age (Kelley 3). Charcoal burns hotter, longer, and with less smoke than wood, and it can be used where wood cannot: in high-temperature hearths, industrial furnaces, and in gunpowder. It is the black in "black powder" that fuels the explosion.

Charcoal has been essential to the production of metal since the Bronze and Iron Ages, beginning 5,500 years ago. Prior to the widespread adoption of coal-derived coke in the nineteenth century, charcoal was the only fuel with enough concentrated energy to fire bloomeries and blast furnaces in ferrous metallurgy. Iron-making technologies supported great advances in human civilization: the iron plow that supported stationary rather than itinerant agriculture, the advances in armor, weaponry, architecture, and machinery enabled by wrought iron, cast iron, and eventually steel. Charcoal can readily drive a fire up to a searing thousand degrees Celsius, a temperature that fuses sand, limestone, and soda ash to form glass. Charred wood is resistant to decay in swampy conditions, so it was used extensively in the foundations of heavy structures built by the Greeks and Romans (Kelley 3). Considering its preeminence in fueling preindustrial metal and glassworks, charcoal was the essential platform fuel between the biomass wood fire in the *Homo erectus* camp more than a million years ago, and the coal-fired furnaces of the Industrial Revolution only two hundred years ago.

Effective charcoal production requires an enduring knowledge tradition. The charcoal-making process essentially speeds up the geological time required to transform forests into coal. It has historically been handmade in isolated pockets of deep woods by colliers who work in long, slow rhythms, enveloped by smoke and indifferent to season. It rewards patience, experience, and artistry in its alchemy from raw biomass to ripe fuel. Talented charcoal colliers can produce their goods with 90 percent efficiency, losing little of the potential energy contained in the wood fiber while smoldering out its water and impurities. Since charcoal weighs only about 15 percent of the raw source wood, it made sense to carry the charcoal long distances, not the wood, particularly when muscles were the movers. The architecture of charcoal work achieves mythic proportions: the smoking pile is a peaked dome with a diameter longer than the standing height of the trees packed inside, and with a height twice or thrice that of its collier. Stacks of cordwood and kindling provide the endoskeleton of an energy hive, and its skin is a sealing mash of turf, clay, and mud.

This giant nest of slow fire is designed to allow the proper minimal circulation of air to the innards of the pile. After first ignition with burning coals, it smokes through the flue at the top like a volcano, and through myriad little holes poked in patterns around the swell of the dome. The collier titers the air holes in

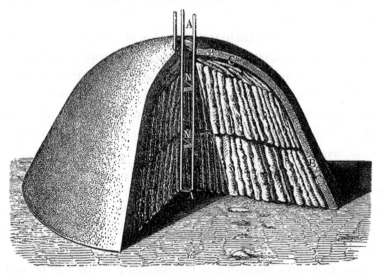

Figure 4.1 Adolf Ledebur, *Meule à charbon de bois*, 1895. Image courtesy of Wikimedia Commons. Public domain.

response to the smoke: blue smoke implies too much oxygen and the collier plugs up some holes; white steamy smoke is perfection—the fire expunges water without burning much of the chemical energy. The collier lives minimally in a conical hut alongside the pile to keep up the conversation over the week or more it takes to make a hill of charcoal from a mountain of wood. Vigilance is key: the wrong kind of combustion, caused by a number of factors like inadequate mud sealing or a breeze buffeting the pile, could quickly ruin many weeks' work. Charcoal burning is traditionally a summer and autumn activity, with the burners becoming woodcutters in winter when trees hold the least sap (Kelley 8). Some colliers settled in one locale and contracted out their goods, some worked nomadically for various woodland landowners. Though charcoal burning has its privations, particularly exposure to the elements, the occupation was important in pre- and early industrial societies and usually paid the skilled burner respectably. Charcoal colliers in England before the First World War earned twenty to thirty shillings a week, a level just below the wages of skilled artisans like smiths (Kelley 8).

Their practice harkens back to Europe between the fourteenth and nineteenth centuries, when iron was required by farmers, blacksmiths, and builders, and forests were stressed to yield enough wood to meet the charcoal demand. An English law of 1558 prohibited the felling of wood for charcoal, citing the nuisance of voluminous smoke by hundreds of colliers, and the fears of scarcity, since wood was also essential for building and ship construction. Since coppice wood was insufficient and England needed iron, this law seems to have been mostly ignored and the forests diminished further. The disparity between wood resources and the fueling of Britain's eighty-five or more blast furnaces, each of which used as much as five tons of charcoal every week, continued through the sixteenth century (Kelley 11).

Some relief to the devastation came during the eighteenth century, when Abraham Darby developed an efficient technique for using coke from coal to replace charcoal in blast furnaces. By 1788, two-thirds of the furnaces in England were fired by coke (Kelley 12). In France, charcoal-fired industries consumed even more forest due to naval and civil building projects. Eighteenth-century French forests were managed and cut at prescribed intervals of ten to fifteen years, which may have made metal-smelting industries reasonably sustainable in that part of Europe at that time (Graham 27–8). Nonetheless, charcoal production and its associated iron-making industries had a ruinous appetite for wood, and the resulting deforestation in Europe and elsewhere wrought major changes at the macro-ecological level, from the Roman-era Mediterranean to seventeenth-century Ireland to modern-day Haiti and Mozambique. Fuel historian Vaclav Smil has remarked on the cumulative pocking effect of charcoal making through history: "When a single furnace could strip each year a circle of forest with a radius of about four kilometers, it was easy to appreciate the cumulative impact of scores of furnaces over a period of many decades" (151). Bob Johnson notes that a single ton of wrought iron required the energy of the annual yield of a twenty-four-acre coppice, a forest land managed to produce the quick growth of small timber (24).

The collier in the European tradition lives a hermetic, esoteric life in literature. He sends inscrutable smoke signals to passersby, interrupting the green cover of deep woods with sudden smoke-filled clearings, working alone with primal fire to create the fuel for bustling urban industries. His work is oxymoronic: rural/industry, smoky/purification, passive/action hermit/entrepreneur. As an independent contractor, he maintains contact in the cities with his buyers, whom he visits at appointed times with the latest haul of energy packed in sacks.

There are surprising similarities between the charcoal collier and the shepherd: both figures work with alternate activity and passivity to assist in the energy transfer from the sun to plants to usable fuel. The shepherd drives his flock to the rich meadows, then sits passively while they graze biomass and make meat. The charcoal collier gathers wood and builds his smoldering hive, then sits passively while it slowly transforms biomass into pseudo-mineral. Both are the conductors of chemical energy conversions with long hours to contemplate. Like shepherding, charcoal burning in the age of fossil fuels seems anachronistic. Coal and coke are cheaper for industrial uses, and their widespread availability from the nineteenth century forward undercut the demand for charcoal.[1]

As an alchemist of fuel, there is an element of wizardry in the collier's work: he feeds forges, intensifies hearth fires, and, since charcoal was a component of gunpowder, fills the barrels of rifles and cannons. There is also an element of idleness in his work; he is necessarily a patient man who watches over a process that completes itself. He is the conductor, quite passive in contrast to the fire's furious activity. Robinson Crusoe, who has all the time in the world to tinker with his abodes, his flock, his food, and his fire, comes upon charcoal by serendipity as he burns wood fires, and discovers its utility as a low-smoke fuel

---

[1] "Charcoal" briquettes of the Kingsford variety commonly used in barbeques are actually stamped nuggets of wood by-products with various chemical additives to aid adhesion and burning. Some barbeque purists still use "lump charcoal" that went through the old conversion process from real wood because it burns hotter, leaves less ash, and foregoes the skulking chemicals that make mass produced pseudo-charcoal so cheap and easy. It costs about eight times more in the American market, so old-school purity comes at a yuppie price.

that does not signal to the cannibals that fill his imagination. His strategy is to process fuel wood in his inland cave, where the smoke would raise less attention, and carry the finished charcoal to his "castle," which is in a pleasanter but more exposed location. For some, the wood smoke signal means safety, but for Crusoe the recluse, it is the clean fire of charcoal that is a promise of security.

The elements of obscurity, method, and filth often combined with the collier's effect on landscape to produce a menacing figure in literature. The large-scale felling of trees to satisfy the keen industrial appetites guaranteed that the charcoal burner was a central figure in environmental concerns about in early modern England. A 1622 poem by P. Hannay records the dual diabolical and racial associations applied to colliers by the residents of Croydon, a town now long engulfed by sprawling London:

> In midst of these stands Croydon, cloth'd in blacke,
>   In a low bottome sinke of all these hills;
> And is receipt of all the durtie wracke,
>   Which from their tops still in abundance trills,
>   The unpav'd lanes with muddie mire it fills
> If one shower falls; or, if that blessing stay,
> You may well smell, but never see your way ...
>
> And those who there inhabit, suting well
>   With such a place, doe either Nigros seeme,
> Or harbingers for Pluto, prince of hell;
>   Or his fire-beaters one might rightly deeme;
>   There sight would make a soule of hell to dreeme,
> Besmear'd with sut, and breathing pitchie smoake,
> Which (save themselves) a living wight would choke.
>
> These, with the demi-gods, still disagreeing
>   (As vice with virtue ever is at jarre)
> With all who in the pleasant woods have being,
>   Doe undertake an everlasting warre,
>   Cut down their groves, and after doe them skarre
> And in a close-pent fire their arbours burne,
> Whileas the muses can doe nought but mourne.—
>
> The other sylvan, with their sight affrighted,
>   Doe flee the place wherease these elves resort,
> Shunning the pleasures which them erst delighted,
>   When they behold these grooms of Pluto's court;

While they doe take their spoiles, and count it sport
To spoil these dainties that them so delighted,
And see them with their ugly shapes affrighted.

(Qtd. in Steinman, 6–7)

This obscure poem powerfully invokes an early modern angst about the evacuation of a virtuous sylvan spirit world at the hand of industrial spoilers. Comparing colliers with Pluto, god of the underworld, Hannay features their filth, blackness, subterranean workings, and materialist motivations in a style that would become featured in environmental literature by the nineteenth century. The poetic muse was shifting from biomass, a mourning sylvan dryad, toward processed industrial fuels with equivocal but beguiling aesthetics, such as the coal that would foment the Romantic vision.

William Blake witnessed firsthand the loss of old-growth, Druid-worshipped oaks on the Surrey Hills south of his home London. The trees were burned by charcoal burners supplying fuel for blast furnaces involved in metallurgy, a crucial industry for weapons supply during the extended war with France (McKusick 99). In his prophetic poem *Milton*, Blake fuses his living experience of tree burning with a vision of global revolution catalyzed by equivocal fiery forces:

The Surrey hills glow like the clinkers of the furnace: Lambeths Vale
Where Jerusalems foundations began: where they were laid in ruins
Where they were laid in ruins from every Nation & Oak Groves rooted
Dark gleams before the Furnace-mouth, a heap of burning ashes.

(Plate 6)

From his vantage point in fast-growing industrial London, Blake viewed history played out on the landscape. In his poem *The Four Zoas* he casts his character Luvah "into the furnaces of affliction & sealed / And Vala fed in cruel delight the furnaces with fire" (Plate 25). Luvah is, in effect, iron that is refined by the blast furnace, "cruel" though necessary work. The sublime of these charcoal furnaces was glowingly visual, and also aural: the air roared as it was forced into the sealed furnace by water bellows, invoking a spirit of suffering (Stevenson 326). Blake was at the leading edge of a literati enthralled with the dark energies of industry, and charcoal is a particularly powerful crux between old wood and new coal, a symbol of biological life sacrificed for the chemical fury of concentrated fuel. The conversion is explicit, unlike underground mines and wells for fossil fuels made of organisms that died in previous epochs. With charcoal, the conversion takes place in open air: forests are thinned and pocked with clearings, smoke wafts a miasma as the wood is purified, and the charcoal burner stands at the center of the sylvan holocaust, meditative by the pyre. Charcoal making is a funeral and a new beginning, a crucial medial element in energy history's paradigm shift from biomass to fossil sources.

Blake eschews the nostalgia of the preindustrial sylvan man, such as the one that Wordsworth alludes to in 1798's "Lines ... Tintern Abbey" in the opening sequence that abuts sublime forms with pastoral ones. The "steep and lofty cliffs / ... on a wild secluded scene" neighbor "plots of cottage-ground" and "pastoral farms / Green to the very door"—cozy farms that reveal themselves within the deep woods by the "wreaths of smoke / Sent up, in silence, from among the trees, / With some uncertain notice, as might seem, / Of vagrant dwellers in the houseless woods, / Or of some hermit's cave, where by his fire / The hermit sits alone" (ll. 5–6, 17–23). Wordsworth's subtle metaphorical shift in the phrase "as might seem" confounds the inviting dwellings, farms and cottages, with the more ominous and open signifier of houseless smoke within the "houseless woods." Vagrants and hermits live on subsistence using their knowledge of the woods, and at the end of the eighteenth century the woods sheltered marginalized people: soldiers dismissed from the Britain's war with France, dispossessed families, religious solitaries, and charcoal burners.

Their fires are poetically "uncertain" to the tourist Wordsworth, serving as objects of mystery and excitement for his meandering contemplation. These hermits contribute to the picturesque intrigue of a forest haunted by unheard stories, textured and scented with idle wafts of smoke. The charcoal collier is one of them; though he is not a refugee from the vicious whims of history and society; he is a recluse by trade. The "uncertain notice" of smoke columns could indicate the idleness of the loitering vagrant or the productivity of the master collier. If the latter, Wordsworth's ambivalence of the smoke brings a surprising industrial element to his lyric masterpiece of Romantic immersion. Nature is all-encompassing, guiding the features even of human constructions, but something mischievous is afoot in the woods. The trees are burning. Characteristically uncurious about the fine details of the scene, Wordsworth stays on the high road and leaves uncertainty to work its faintly repellant force on the reader's mind.

His fire-breathing, sooty aspect and dwelling on the margins of society do not always make the collier a menacing figure. He lives happily in children's literature, for example in A. A. Milne's short lyric in his 1927 book *Now We Are Six*:

The charcoal-burner has tales to tell.
He lives in the Forest,
Alone in the Forest;

He sits in the Forest,
Alone in the Forest.
And the sun comes slanting between the trees,
And rabbits come up, and they give him good-morning,
And rabbits come up and say, "Beautiful morning ...."
And the moon swings clear of the tall black trees,
And owls fly over and wish him good-night,
Quietly over to wish him good-night ....
And he sits and thinks of the things they know,
He and the Forest, alone together—
The springs that come and the summers that go,
Autumn dew on bracken and heather,
The drip of the Forest beneath the snow ....
All the things they have seen,
All the things they have heard:
An April sky swept clean and the song of a bird ...
Oh, the charcoal-burner has tales to tell!
And he lives in the Forest and knows us well.

(ll. 1–21)

Milne's collier is the natural man, living outdoors through the seasons with a lively society of his animal neighbors. Rather than the fell deforester, he is a sylvan in harmony with the forest ecosystem and becomes its interpreter and translator; he looks, he thinks, he tells. He has a better celestial perspective because of the clearing where he burns; the opening surrounded by tall black trees frames the sun and moon. The seasons efface the evidence of burning by way of snowy drips and spring breezes, so the natural architecture of his space is lofty and clean, rather than low, secretive, and grimed. The charcoal burner has tales to tell to children because, like Snow White, his friends are the humanoid animals that town-bound people lack the patience and perception to know. His gift, aside from the charcoal itself, is the susceptibility to small signs of beauty and gentle dwelling. Where Wordsworth's "vagrant dwellers" carry a repulsive anxiety, Milne's poem ushers children toward the charcoal burner, ambassador of the deep woods. Milne's nostalgia shows the obsolescence of the charcoal burner and his rural dwelling, eclipsed by giant urban industries fired first by coke and coal, then by natural gas and electricity. By the nineteenth century, the charcoal burner was an endangered species in the European woods, and the forge fires were fossil-fueled.

John McPhee describes charcoal burning in the New Jersey Pine Barrens as a major occupation until the Second World War. The first European people to occupy the Barrens were seventeenth-century woodcutters who founded a charcoal industry for the growing cities east of Appalachia. Their best charcoal was hauled by eight-horse teams across the sands specifically to fire the Philadelphia mint. Much of the rest was transported in dozens of schooners up the coast to New York City (45). Pine Barrens charcoal fueled a major industry in pig iron that carried the local economy for decades during the late eighteenth and early nineteenth centuries. Like the whaling industry, and at about the same time, the charcoal industry went into decline in the mid-1800s when raw supplies grew scarce and, perhaps more importantly, coal from western Pennsylvania emerged as a cheap and functionally infinite supply of fuel. Coal and its derivative, kerosene, replaced, respectively, charcoal and whale oil; it was cheaper as both industrial fuel and as illuminant. Within twenty years, the major furnaces of the Barrens—with names like Batisto, Martha, and Atsio—were closed, as the land could no longer supply the thousand acres of woods per year it took to fire them. Charcoal burning shrank down to a small local business again (38).

Thereafter charcoal burning in the Pine Barrens remained one occupation among others, as part of a local economy that included the gathering of sphagnum moss in the early spring (a natural sponge, decoration, and antiseptic), wild blueberries in the summer, cranberries in the fall, and cordwood and charcoal in winter (43–4). Charcoal burners were known to set forest fires so as to damage trees and make them good for nothing but charcoal, thus they could be purchased at a discount price (115). The Barrens bear the scars of their nearly four-hundred-year investment in charcoal burning: the area is pocked with pit clearings that form cul-de-sacs at the ends of sand roads, a maze-like geography that has flummoxed many a visitor. Now that charcoal burning is virtually obsolete in America, these ghost pits evanesce into closed forest as the years pass.

The United States had the advantage over Europe of being able to preserve vast tracts of forest through the adoption of coal; western Europe had endured centuries of forest stress from the manufacture of charcoal, among other uses, before coal assumed the burden of firing metallurgy. In retrospect, conversion to coal in the United States saved what remained of the mighty, diverse forested lands, and may deserve credit for the preservation of wilderness within the monumental park system (Johnson 23). The ancient, subterranean forest saved the extant, solar one.

Small industries involving charcoal endure as industrial subcultures to this day. The Italian documentary film *La Quattro Volte*, "The four turns," ends on an extended sequence of the collier's process, from the harvesting and arrangement of the burning pile, to its maintenance, and finally to the delivery of sacks of charcoal to picturesque whitewashed doorways in hilly Calabria. From there, the viewer imagines, it will fire home ovens that roast tomatoes and brown focaccia. The only element in the film that is not medieval is the charcoal burners' use of a tiny truck cart rather than a beast of burden. The fifteen-minute sequence is unnarrated, contemplative, and closely focused, transporting the (urban industrial technocrat) viewer to a premodern world characterized by slow timeworn traditions of knowledge and basic subsistence. It says, this is fuel, and fuel is more than a gas tank or an outlet: it is a small sack of black rubble in a doorway; it is the afterlife of an Italian forest. Small-scale charcoal burners like these continue to supply markets throughout the developing world.

II Selling charcoal

After the fuel was produced in the forest, colliers then became salesmen in the city. Singing the toils of the plebian charcoal seller, the American writer John Townsend Trowbridge, a friend of Mark Twain and Walt Whitman, published a poem in 1869 on "The Charcoal Man." Trowbridge walks the nostalgic path seen in Milne's children's poem, with a conventional good soul concealed beneath layers of honest work, whose toil is duly rewarded. Trowbridge's scene depicts the urban side of the business: rather than conversing with nature, this charcoal man sings the call-and-response of the northern industrial city.

Mark Haley drives his cart through snowy streets bellowing "Charco'! Charco'!" while the city sings back "Hark, O!" Street boys mock his cart with "Ark, ho!" his wife greets him with "Mark, ho!" and even his baby responds with "Ah-go!" The cheery music ringing around the besmirched roughneck is in harmony with his inner purity, and Trowbridge sings a populist message: "Thus all the cold and wintry day / He labors much for little pay; / Yet feels no less of happiness / Than many a richer man, I guess / ... The hearth is warm, the fire is bright / And while his hand, washed clean and white, / Holds Martha's tender hand once more, / His glowing face bends fondly o'er / The crib wherein his darling lies" (ll. 23–7, 44–8).

Hale Mark embodies the right honest tradesman patient of his labor and its hardships, rewarded with a faithful loving wife, a cooing baby, and, thanks to his surplus, an especially warm and bright fire. Trowbridge's (at best) simplistic moral is to honor the romanticized poor man: "Though dusky as an African, / 'Tis not for you, that chance to be / A little better clad than he, / His honest manhood to despise" (ll. 46–9). Trowbridge's idealism seems to be free of the irony that marks Blake's "Chimney Sweeper" from the Songs of Innocence, who similarly cries "'Weep! weep!'" and lives in racially ambiguous skin because of his sooty trade. Both figures are peons, but Blake's boy, like his slaving African counterparts, questions his status and worries about his physical and spiritual health.

Mark Haley is above disdain because he is not truly "dusky as an African"; his hands are "washed clean and white" and his manhood is "honest" (whatever that means). The implicit racism here is derivative of centuries-old imagery rather than a forward-looking post–Civil War fresh leaf. It shows the indelible stigma of dirty work in an age turning toward white-collar business, law, and stock speculation that would within a decade become the Gilded Age in America. Mark Twain and Charles Dudley Warner coined the term "Gilded Age" in their 1873 satire of that period of corrosive general poverty, thinly covered by the gold gilding that signified the extremely wealthy few and their conspicuous consumption. (Edith Wharton and Henry James would make a career in gilded portraits.) Trowbridge seemingly begs the oblivious reader to acknowledge the existence of this dirty, honest, happy, simple fellow who fills the hearths of the city. However, Mark Haley's repetitive music is the jingle of the trite commercial. Milne's charcoal-burner has wisdom and keeps the secrets of the woods, but Mark, the city charcoal man, is allowed no inner life. His book is just its cover, illustrated in the Rockwell style.

An intriguing counterpoint to this charcoal man is a holler of the same name by the journalist Lafcadio Hearn that dates from the 1880s. Hearn was a newspaper correspondent based in New Orleans who published accounts of the Creole people, their cuisine, and their voodoo practices. Just like Trowbridge, Hearn remembers the charcoal man by his oral presence in the city streets, and the poems both read as joyous music that lightens the load of those who barely subsist by selling charcoal. Where Mark Haley sang in a tight loop ("Charco'! Charco'!"), Hearn's Creole crier continually innovates his song:

Char-coal, Lady! Char-coal! Chah-ah-coal, Lady!
Black—coalee—coalee!
Coaly—coaly; coaly-coaly—coal—coal—coal.
Coaly—coaly!
Coal—eee! Nice!
Chah—coal!
Twenty-five! Whew!
O Charco-oh-oh-oh-h-oh-lee!

Oh—lee—eee!
(You get some coal in your mout', young fellow, if you don't keep it shut!)
Pretty coalee—oh—lee!
Charcoal!

(ll. 1–12)

Hearn celebrates the crier breaking the constraints of grammar and syntax, finding a form of infinity within repetition that continually creates new sounds within and outside known language. Romanticists would recognize this revolutionary music in the poetry of William Blake, who sought to unlock language from its repressive old-world fetters or, as William Wordsworth said, to break out of a world "cased in the unfeeling armour of old time" (Rothenberg and Robinson 707). Hearn's found poem is experimental, improvised, and performed—it is a predecessor of slam poetry—and its free verse arrives at a free place, transcending the formal conventions of blank verse, sonnets, and Trowbridge's tetrameter couplets. It looks to the future, to the twentieth century, when actual sound recordings of these songs became possible, and it looks to cultures beyond the Anglophone hegemony of American culture. The charcoal man continues:

Charbon! Du charbon, Madame! Bon charbon? Point! Ai-ai!
Tonnère de dieu!
Cha-r-r-r-r-r-rbon!
A-a-a-a-a-a-aw!
Vingt-cinq! Nice coalee! Coalee!
Coaly-coal-coal!
Pretty Coaly!
Charbon de Paris!
De Paris, Madame; de Paris!

(ll. 14–22)

As he slips from English into a similar riff in French, the charcoal man yells out his punctuation ("*Point!*" indicates full stop or an exclamation point), and calls in God's thunder, "*Tonnère de dieu!*" as another exclamation or invocation of the power of the fuel he sells. At twenty-five ("*Vingt-cinq!*"), an unknown volume in undeclared currency, the holler is both precise and elusive. The charcoal man's pitches are aimed at women who tend the domestic fires of the city and make the essential purchases, so his song is also a flirtation, wherein the charcoal takes on the features of his buyers—black, pretty, nice. The humble charcoal, the "*Charbon de Paris*," becomes exotic by its association with the City of Lights: stylish Paris flickers in the imaginations of Creole New Orleans, which empties its pockets in response to the call.

In both of these thoroughly different charcoal poems, the singing salesman sketches an aural cityscape in a call-and-response, where hecklers join in the song (both men sass roguish youths), and implied buyers may respond to the holler with their clanking coins. But it is mostly a soliloquy of charcoal, the hot-burning wood of the urban hearth. These are ghostly cries that haunt our cities, along with the songs of ice and milk that were commonplace before the advent of electricity. A few decades into the twentieth century, once most houses possessed a means to fire and freeze via electricity, the salesmen faded away and the fuel, most often coal, burned obscurely in electric plant furnaces. To the modern consciousness, fuel became an electric bill, and its performances vanished.

III Blow it up: Gunpowder and fireworks

Every musket and cannon fired in every war before the nineteenth century, every rifle shouldered by a hunter, every firework blooming over celebrations, owes its bang to charcoal. It was the only workable incendiary available before coke, from coal, became widespread in the nineteenth century. Charcoal, sulfur, and saltpeter (potassium nitrate) are the three ingredients of basic gunpowder. The charcoal used for gunpowder needs a bit more care in preparation: the alder and willow trees most commonly used as the raw material had to be cleaned of all bark, and all the grit had to be hand removed after carbonization, before the charcoal could proceed to the powder works (Kelley 10–1). English soldiers first used gunpowder in 1346 at the Battle of Crecy, where they were outnumbered two-to-one, but the English cannons and crossbowman made for heavy losses on the French side, and an English victory (Kelley 10). Guy Fawkes owes his death and immortality to gunpowder that never caught fire, though since then fireworks have exploded every November 5 in his dubious honor.

William Shakespeare and John Milton witnessed the execution of Guy Fawkes and his conspirators in 1605. Some argue that Shakespeare's *Macbeth* alludes to the Gunpowder Plot; Milton is more explicit in *Paradise Lost*, where, in book six, Satan and his minions make the machines of modern warfare— cannons and gunpowder—to battle with God's angels. Satan does not come to his dastardly machinations lightly. He has been gravely injured by Michael's two-handed sword of Divine Justice, which "deep ent'ring shear'd / All his right side" (6.326–7). Satan finds that "pain is perfect misery, the worst / Of evils, and excessive, overturns / All patience," and thus is inspired to destructive cleverness (6.462–4). To the rebel angels, he observes the beauty of heaven, "With Plant, Fruit, Flow'r Ambrosial, Gems and Gold," and notes that these pleasant dressings conceal a more potent cache of natural wealth underground. Transferring his imagination from heaven to the hell he would soon occupy, Satan describes

> materials dark and crude,
> Of spirituous and fiery spume, till toucht
> With Heav'n's ray, and temper'd they shoot forth
> So beauteous, op'ning to the ambient light.
> These in thir dark Nativity the Deep
> Shall yield us, pregnant with infernal flame,
> Which into hollow Engines long and round
> Thick ramm'd, at th' other bore with touch of fire
> Dilated and infuriate shall send forth
> From far with thund'ring noise among our foes
> Such implements of mischief as shall dash
> To pieces, and o'erwhelm whatever stands
> Adverse, that they shall fear we have disarm'd
> The Thunderer of his only dreaded bolt.
> Not long shall be our labor, yet ere dawn,
> Effect shall end our wish.
>
> (6.478–92)

Satan's science, geology and pyro-chemistry, are figured as the evil contrivances of a cunning devil. This cunning contrast the pure justice of the right angels, who work with simple weapons (swords and arrows), blessed by God's grace. Explosives are not only naughty, they are also the arms of the lazy. "Not long shall be our labor," Satan opines, as they dig:

> Sulphurous and Nitrous Foam
> They found, they mingl'd, and with subtle Art,
> Concocted and adusted they reduc'd
> To blackest grain, and into store convey'd:
> Part hidd'n veins digg'd up (nor hath this Earth
> Entrails unlike) of Mineral and Stone,
> Whereof to found thir Engines and thir Balls
> Of missive ruin; part incentive reed
> Provide, pernicious with one touch of fire.
>
> (6.512–20)

Mixing volatile and smelly compounds, the work is dirty, "concocted" or cooked, and "adusted"—a medical term that meant that the compounds were finely ground (Hughes, note p. 335–6). Satan is a cunning chemist who mixes burnt trees with the minerals sulfur and saltpeter in order to make gunpowder. The rebel angels are duly impressed with this weaponized fuel at their disposal. They hasten to the next day's battle and put fire to the fuse:

> Immediate in a flame,
> But soon obscur'd with smoke, all Heav'n appear'd,
> From those deep-throated Engines belcht, whose roar
> Embowell'd with outrageous noise the Air,
> And all her entrails tore, disgorging foul
> Thir devilish glut, chain'd Thunderbolts and Hail
> Of Iron Globes, which on the Victor Host
> Levell'd, with such impetuous fury smote,
> That whom they hit, none on thir feet might stand …
> Foul dissipation follow'd and forc't rout.

(6.584–92, 598)

Milton portrays this nasty warfare using body-based descriptions of throats, bowels, and entrails. He figures the new battle tactic as a disgraceful, vomiting, diarrhea-cursed illness. *Real* angels would never fight in this felonious way. These Satan-angels are dirty birds, suspect in their cunning, sinful in their concoctions. Still, Michael's pure two-handed sword needs upgrading, so the right angels regroup, join the arms race, and start their own devastating counterattack. They throw down the "Rocks, Waters, Woods" and the very mountains on the rebels. God approves of this zealous reply, which nevertheless results in heaven "gone to wrack, with ruin overspread" (ll. 645, 670). The scorched earth warfare allows his son, the Messiah, to condemn the Satanists to hell, and take his seat "at the right hand of bliss" (l. 892).

Quite unlike the gentle wise forest collier, Milton's memorable contribution to the literature of gunpowder points to the ruinous efficacy of charcoal chemistry. Gunpowder is a weapon in the hands of the devious clever man, both archetypally and literally Satan in this case, whose schemes challenge but fail to overthrow the imperious God. Satan's character was admired by the British Romantics as embodying, in Coleridge's words, the "dark and savage grandeur" so often seen in historical despots from "Nimrod to Bonaparte" (364). To the Romantics, Satan is admirable in his revolutionary zeal, his singular purpose, and the rebellious personality that enables such stunning insubordination without regard to the means employed. His temerity puts the justice of God's autocracy into question.

He helps his case with eloquent soliloquies that reveal his ambition and his agony. The British Romantics saw the value of an unscrupulous revolutionary willing to take on a despotic ruler (for them, Kings George III and IV), but they also admired the audacity of Satan's methods. Coleridge names his "outward restlessness and whirling activity; [his] violence with guile" as elements of an admirable character (153). Percy Shelley, declaring that Satan is ennobled as the loser of the war in heaven, finds that his weapon, the "sleepless refinement of device to inflict the extremest anguish on an enemy" is "venial in a slave" (127). However, the gunpowder plot is pardonable for Shelley because God is a tyrant, and so the revolution is just.

In the century following the publication of Milton's classic, Daniel Defoe would cite gunpowder as an element of basic survival in *Robinson Crusoe*, when the castaway manages to recover a wet cask of powder from his foundered ship. Crusoe has many guns that are utterly useless without the powder, so he is delighted to discover, in his twenty-third year of solitude, a large cache of gunpowder in a cask that had been soaked during the shipwreck. Crusoe reports: "This was a very agreeable discovery to me at that time; so I carried all away thither, never keeping above two or three pounds of powder with me in my castle, for fear of a surprise of any kind; I also carried thither all the lead I had left for bullets" (140). Crusoe lives in continual fear of cannibals visiting the island, so the powder has less of practical applied use than it does as a psychological buoy for his schemes to blow the "savages" away. Friday's arrival complicates Crusoe's prejudices, but the powder remains fodder for Crusoe's bloody fantasies, though never a reality.

Soon after Defoe, Jonathan Swift would deploy gunpowder as an element of an advanced and powerful civilization—not the tool of the revolutionary, but of the hegemon. In *Gulliver's Travels*, the hero repeatedly astonishes neophytes, tiny and towering, with the pow of gunpowder. It lends him the luster of superhuman agency regardless of his relative size to the people he visits. It sparks Swift's searing satire of the ubiquitous violence and warfare of his day due to superficial religious and cultural differences. Swift's explosive descriptions also remind the reader of the dreaded destructive potential of gunpowder, which had become mundane due to its common military use beginning in the fourteenth century. The wee Lilliputians search him and find his pistol, which, when shot, causes hundreds of the tiny people to fall down as if "they had been struck dead; and even the emperor, although he stood his ground, could not recover himself for some time" (45). Gulliver advises the emperor of Lilliput to spare the pouch of powder from his fire, "for it would kindle with the smallest spark, and blow up his imperial palace into the air" (45). Like Swift's countrymen, the Lilliputians are combative, mechanical, profit-oriented, and full of admiration for the ruinous potential of gunpowder.

Gulliver is homesick during his stay in Brobdignag, the land of the giants. He finds the relative isolation of the Brobdignagians the primary cause of their "narrowness of thinking" and, to show "the miserable effects of a confined education," he regales the king with the wonders of gunpowder:

> I told him of an invention, discovered between three and four hundred years ago, to make a certain powder, into a heap of which, the smallest spark of fire falling, would kindle the whole in a moment, although it were as big as a mountain, and make it all fly up in the air together, with a noise and agitation greater than thunder. That a proper quantity of this powder rammed into a hollow tube of brass or iron, according to its bigness, would drive a ball of iron or lead, with such violence and speed, as nothing was able to sustain its force. That the largest balls thus discharged, would not only destroy whole ranks of an army at once, but batter the strongest walls to the ground, sink down ships, with a thousand men in each, to the bottom of the sea, and when linked together by a chain, would cut through masts and rigging, divide hundreds of bodies in the middle, and lay all waste before them. That we often put this powder into large hollow balls of iron, and discharged them by an engine into some city we were besieging, which would rip up the pavements, tear the houses to pieces, burst and throw splinters on every side, dashing out the brains of all who came near. That I knew the ingredients very well, which were cheap and common; I understood the manner of compounding them, and could direct his workmen how to make those tubes, of a size proportionable to all other things in his majesty's kingdom, and the largest need not be above a hundred feet long; twenty or thirty of which tubes, charged with the proper quantity of powder and balls, would batter down the walls of the strongest town in his dominions in a few hours, or destroy the whole metropolis, if ever it should pretend to dispute his absolute commands. (161–2)

Gulliver openly delights in the melee, for how could a proud Englishman explain the defeat of the Spanish Armada or the Battle of Blenheim without reference to the almighty cannon? The king of Brobdignag, however, is horrified by the "terrible engines" of destruction Gulliver describes, and lashes Gulliver—an "impotent and groveling insect"—with his disdain for the Europeans' inhuman weapons (162). Brobdignag is a land far removed from the endless bloody squabbling of Europe, located somewhere in the American Pacific northwest, and it is governed by the values of sense, justice, and simplistic reason. Like most utopias, Brobdignag is horribly boring. Gulliver would spark no fireworks there. The obscenity of eighteenth-century warfare becomes a leitmotif in the travels, and Gulliver's avid descriptions of munitions warfare repeatedly offend his audience. To the equine Houyhnhnms, who take pride in their hooves and teeth as tools of destruction, Gulliver describes the rumpus of English battle:

> Being no stranger to the art of war, I gave him a description of cannons, culverins, muskets, carabines, pistols, bullets, powder, swords, bayonets, battles, sieges, retreats, attacks, undermines, countermines, bombardments, sea fights, ships sunk with a thousand men, twenty thousand killed on each side, dying groans, limbs flying in the air, smoke, noise, confusion, trampling to death under horses' feet, flight, pursuit, victory; fields strewed with carcases, left for food to dogs and wolves and birds of prey; plundering, stripping, ravishing, burning, and destroying. And to set forth the valour of my own dear countrymen, I assured him, "that I had seen them blow up a hundred enemies at once in a siege, and as many in a ship, and beheld the dead bodies drop down in pieces from the clouds, to the great diversion of the spectators." (291–2)

Observing the dubious conflation of violence with fun, Gulliver's audiences can hardly disguise their moral repulsion as he puffs the power of the powder, testing their admiration of English values. To the wise Houyhnhnm, the ability of humans simultaneously to possess both reason and bloodlust is evidence of moral corruption; the Houyhnhnm concludes that rather than reason, humans disclose the capacity for clever cunning, which inflates and distorts their self-image and drives them to build such sad, horrific, weapons. No such warfare would be permissible, let alone celebrated, in a true age of reason. In *Gulliver's Travels*, the little men are violent and querulous, the big men are passive and imperious, and the horses are wise beyond all men. The position on gunpowder is a moral litmus test in each of the visited worlds.

One last excursion into charcoal-fueled literature comes from Russell Hoban's 1980 novel *Riddley Walker*. Nuclear warfare has cast civilization into a neo-Iron Age a few thousand years in the future. Hoban speculates on the renewed value of pre-fossil-fuel sources and the preserved wisdom of cultish charcoal burners who are, in effect, skilled workmen performing a crucial task in a fuel-starved world. The charcoal burners are alchemists who rediscover gunpowder, in effect repairing the historical loop that had been torn by nuclear apocalypse. Once the future-primitive world has gunpowder, it is back in the arms race that hastens inevitably to muskets, heavy munitions, and eventually the bomb.

With human belligerence still alive and well, the society set in southeast England consists of a ring of rival bands racing to rediscover weapons as they scrape out survival in a dismal ecological dystopia. The culture is tied together with a series of oral myths transmitted through doggerel poetry and, farcically, a Punch and Judy roadshow that preserves the knowledge of the bygone industrial civilization that self-destructed by detonating the nuclear "1 Big 1." Charcoal is the heart of the wood ("hart of the wud"); it lies at the center of the mythology and oral history of Riddley Walker's society. Charcoal burners manage coppices of alder trees for a raw fuel source, which they cut at six-year intervals. Alder is an appropriate tree not only because it produces high-quality charcoal, but also because it is a fast-growing species that grows on marshy ground and can tolerate the tainted low-nutrient soil that carpets this nuclear winter-scape. The burners dress ritually in cloth dyed red from the sap of cut alder branches. They are comparatively well fed and safe because they are fuel providers, but the charcoal burners have greater ambitions than to warm hearths.

Riddley Walker is written in a decipherable post-English dialect. A charcoal burner, Granser, says:

> There wont be no 1 Big 1 for us Abel we aint got the cleverness for it but befor there ben the 1 Big 1 there ben the 1 Littl 1 and wil be agen. Which weare waiting for the day and ready ... Ready with 2 of the 3 aint we and ben ready this long time and keeping ready and til it comes never mynd how long it takes the chard coal berners wil be wearing red and keeping ready. (188)

The "berners" hold the knowledge of how to make gunpowder, so the "1 Littl 1" will be possible again once the third ingredient—sulfur—is discovered. They call it "the yellerboy stoan." They already have saltpeter ("Saul & Peter") and "chard coal," so they keep a vigil built on alchemy and superstition waiting for this yellow stone. Without knowing its significance, Riddley Walker has rediscovered it in the coastal town Fork Stone (Folkestone), and brought it back to the charcoal burner's roost at The Auders, near Cambry (Canterbury), which was ground zero for the nuclear explosion of yore. Riddley delivers the yellow stone to Granser:

> He wer pounding the yellerboy stoan to a fine powder. Then he done the same with some chard coal. Done it with a boal and pounder. He had the Saul & Peter all ready that were kirstels like salt. He took little measurs and measuring out yellerboy and chard coal and Saul & Peter. Mixing them all to gether then and me watching. It wer like the 1st time I seen a woman open for me and I wer thinking: This is what its all about then. Granser singing to his self wylst he mixt the 3 of the 1:

Wewl make such a noys
Such a noys

Such a noys
Wewl make such a noys
My good old boys

(193)

This esoteric chemistry is erotic to Riddley because it is a sacred union of diverse elements, the essence of power–pleasure in an enervated world. Granser sends Riddley away to the river as he performs the final incantation, the "fissional seakerts of the act," and then

> WHAP! there come like a thunner clap it wer like when litening strikes right close it eckowit up and down the rivver. There come up a cloud of smoak from the fents it wernt the regler blue smoak it were 1 big puff of grey smoak and things wer peltering down out of the trees like when you shake down nuts.... Granser he wer like throwt a way on the groun he lookit emty like when you take your han out of a show figger. His head wernt with the res of him his head wer on a poal. The gate a way from the river it had a hy poal on each side of it and Gransers head wer stuck on the point of 1 of them poals jus like it ben put there for telling. (194)

Granser's head poling is an omen of the hazard of being "clevver"—the kind of thinking that in the past led to nuclear war and the near extinction of humanity. The gunpowder is on a continuum with nuclear fusion, morally equivalent and only distinguished by the sophistication of the society that supports the research. In all of these gunpowder texts, the use of fuel to destroy is central to moral indictments against any society that squanders energy in this way. Gunpowder foreshadows the appearance of nuclear energy in literature; in both cases, the cautionary principle is strong against a military hegemony.

Early modern Europeans found a nonviolent and diverting outlet for charcoal in fireworks displays. In early modern technical manuscripts, the machinery of warfare and fiery spectacles for public occasions were often similar in design, though functionally antithetical. The sixteenth-century Italian metallurgist Vannoccio Biringuccio noted that "artificial fires" have a "happy and pleasing effect" despite their potential for "harm and terror" and effects that might become "deadly and very injurious to all living things" (Wolfe 236–7). Natural philosophers, chemists, and artisans with applied handwork skills collaborated to take a Chinese invention of the seventh century and make it better: brighter, more colorful, bursting higher, and taking symbolic shapes. Early fireworks were used only for warfare, but beginning in the late sixteenth century there was a flurry of investment in pyrotechnics in England, France, Italy, and Russia that aimed to cash in on the pleasure, edification, symbolism, and patriotism that fireworks might instill (Werrett 10–16).

The first fireworks evoked the sun, moon, and planets, meant as special effects in the theatre of the universe. Accounts of fireworks tended to distinguish the calm masterful appreciation of the courtiers from the superstitious horror of the commoners, a classist spin on the fact that fireworks have always had the sublime power to inspire both fear and delight. Firework displays suggested models of the cosmos and nature to scientists like Isaac Newton, Johannes Kepler, and Robert Boyle, and provided zesty food for thought for philosophers Gottfried Leibniz and Rene Descartes (Werrett 12). Under the purview of national academies, they became "rational recreations" deployed to educate the middle classes about natural philosophy (Werrett 15). In absolutist states, fireworks were used to buoy the authority of the monarch, as when French sycophants wrote that the element of fire was under the control of the miraculous King Louis XIV (Werrett 14). They ingratiated the squeamish masses to Peter the Great's military ambitions by making the violence of munitions warfare more palatable and playful for the people (Werrett 13). Chemistry, a fashionable science, helped differentiate colors in fireworks, producing stunning reds, yellows, greens, and blues by experimenting with the effects of various volatile metals, salts, and charcoals. When electricity became the object of universal enthrallment in the eighteenth century, fireworks inspired some of the first (inaccurate) theories on the nature of "electric fire," based on the idea that electric sparks were pyrotechnic and that thunder and lightning involved combustion (Werrett 16).

James Boswell had "little doubt" that Dr. Johnson had written the 1749 article in *Gentleman's Magazine* that described the chemistry, manufacture, and aesthetic philosophy behind these sublime "sky-rockets." The challenge, the ghost writer explains, is to tightly pack the gunpowder so that it burns rather slowly rather than exploding all at once. Also "by varying the proportions of the brimstone and charcoal, not only the duration of the flame, but its colour and appearance may be considerably changed" (56). The artisan increases the proportion of "grossly powdered" charcoal to brimstone when he desires "an extensive train of fire." Johnson favors the rocket as the "most ingenious, and (if the artificial sun be excepted) the most beautiful of all pyrotechnical representations," and he goes on to describe its aesthetic merit based on the "largeness of the train of fire it emits, the solemnity of its motion (which should be rather slow at first, but augmenting as it rises) the straightness of its height, and the height to which it ascends" (56). Johnson estimates this height to be about 1,500 feet, but he encourages "all gentlemen of leisure and curiosity" to experiment with the composition of rockets and measure the heights they can attain, which "might furnish us with a more certain theory of those motions . . . and would be attended with many practical and philosophical advantages" (56). The English seldom let pass an opportunity to smother childish fun with practical erudition.

In Johnson's mind, the only rival to the splendor of the rocket is the pyrotechnical sun that was displayed at the Triumphal Arch in Green Park in April 1749. The fifty-foot diameter fire-sun, with 180 cases of incendiary mixed with steel dust (a brightener), arranged in a circle, "will all be fired at the same instant, they will compose a body of the most brilliant light . . . will be strongly distinguished with the appearance of rays, after the manner in which the glory round the heads of saints is usually drawn by painters, and will on the whole, afford the grandest and most striking appearance hitherto invented by human art" (56). In true Enlightenment hubris, Johnson concludes, with a touch of sarcasm, that the sublimity of the pyrotechnic sun threatens to outdo the glory of the real thing. Alas, in the cold light beyond human egocentrism, a packet of charcoal, saltpeter, and sulfur ultimately pales in comparison with the sun, a spherical storm of hot plasma and magnetic fields 330,000 times the mass of earth.

Fireworks continue to provide the mass ceremonial spectacle that they have for hundreds of years. They are skyborne incendiary rhetoric that still reinforces hegemony, especially as they are deployed in Independence Day celebrations and baseball games. They fuse power with beauty, and danger with technical control. Instilled as they are in global culture, fireworks displays now get the most attention when they go wrong or cause problems. On July 4, 2012 the city of San Diego's entire seventeen-minute show exploded within one blinding, deafening minute due to a corrupted file in a computer program. On the Chinese New Year day in 2014, fireworks were suppressed so that their burning didn't contribute to the already oppressive coal smoke and petroleum exhaust that plagues today's Chinese cities. The editors of *China Daily* newspaper said "it should be common sense" that revelers celebrate the new year "without coughing and spluttering on the after-effects of fireworks" (Rochan). With Chinese meteorologists joining in the anti-firework campaign, the Beijing Consumer Association recommended flowers as a substitute. Their suppression in China is particularly poignant: the Chinese discovered fireworks in the seventh century, long before they were known in Europe, and quickly adapted the displays to ceremonial and symbolic uses. Their loss is an indication of the power of fossil fuels, especially coal, to dominate a culture's pollution quota in the name of industrial production.

As it was in eighteenth-century Europe and nineteenth-century America, charcoal production is a modern environmental concern in developing countries, where charcoal remains a primary fuel source. A Greenpeace investigation into the production of charcoal in the Amazon makes it relevant to industrial consumers ("Driving Destruction in the Amazon"). Deforestation in the vast Carajás region of Brazil is driven by the global demand for pig iron, an ingredient of steel in the supply chain of several car manufacturers: Greenpeace names Ford, General Motors, BMW, Mercedes, Nissan, and John Deere (3). Remote charcoal camps supply fuel to the world's largest iron mine, which produces pig iron from raw iron ore. (The ore mine itself is mostly fueled hydroelectrically by the Tocantins River via the Tucurí Dam.) The charcoal camps can be moved rapidly to avoid detection and regulation, and to exploit fresh supplies of raw timber from rainforests, often on protected or indigenous lands. Economics drives the industry toward illegal logging: rainforest timber costs about one-tenth the cost of plantation-grown timber, and the charcoal that fuels pig iron furnaces represents as much as 35 percent of the cost of iron production (5). It is estimated that one-third of the timber felled for charcoal production in Carajás is illegally harvested, and the cleared land usually becomes pasture and soy farms rather than reverting to forest (8).

The case here illustrates the complex relationship between numerous fuel sources used to produce goods in the global economy. The sun (renewable) energizes trees (raw biomass) that are harvested for charcoal (processed biomass); together with raw iron ore (mined using renewable hydropower and fossil fuels), charcoal fuels the production of pig iron, which is shipped using petroleum bunker fuel (fossil) to steel manufacturers, who fire their furnaces with coke derived from coal (fossil). If the end product of the steel is a car, it rolls around fueled by, and partially composed of, petroleum (fossil). The wood grain interior in a fine Mercedes is a ghostly shadow of Amazonia, burned down to provide the steel in the car body.

Wood fuel is the original source of the controlled human fire. Its rituals are stamped on our DNA and our phenotype, adapted as we are to chew and digest cooked foods. Its critical importance in human prehistory, the radiation out of Africa, the energy to clear land for agriculture and other exploits, the courage to traverse wintery expanses cradling hot coals, makes the wood fire the most evocative of preindustrial fuels. The physical subsistence offered by the wood fire goes in hand with the cultural significance of burning wood, rituals of seasonal preparation and survival, appearances of wealth and resourcefulness, fears of shortfalls. Grass and fodder are the all-important animal fuels that have supported hunting and agricultural groups in the form of perennial cud. But regal wood is the human incarnation of biomass energy.

Wood's afterlife in charcoal greatly expands the potential of the sylvan trope. As the primary segue fuel for the industrial era, a premonition of the power of coal and gas, charcoal commands an amazing variety of roles. From the home fires of New Orleans Creoles to the blast furnaces of early industrial London, from plate mail and steel swords to iron bridges and delicate glass domes, from the fury of the cannon to the fantasy of the firework, charcoal is both diabolical and angelic. As Riddley Walker knows, it is the heart of the wood.

Part Two

Fossil Fuels

5

King Coal

The paradigm shift is at hand. Earth's most abundant fossil fuel, coal, powered the dynamos of the Industrial Revolution beginning in the eighteenth century. The union of coal with steam engines changed the world. Their synergy required mutual, enabling innovations between the mining and mechanical sciences, which developed rapidly over the eighteenth and nineteenth centuries in Europe and North America. In the wake of this union of fuel with machine, global ecology and human society shifted into a new industrial era. Natural resources conserved by the limits of a low-energy world could be exploited on much larger scales, and the agricultural labor required to produce goods like cotton, which supplied textile mills in England, grew proportionally.

Instead of transportation guided by natural conditions like seasons and weather, coal-fired engines worked human society into strict timetables, largely indifferent to adversity. Our conception of distance expanded from local foot-and-horse reach to regional and global networks of trade and leisure travel. The hard physical work required to produce raw goods, which tended to pay poorly or not at all, continued apace, while the owners of the new industrial architecture shifted to cognitive and managerial roles: the division between blue- and white-collar work. Fueled by coal, a new high-energy economy fostered the capitalist mentality of competition and innovation, and also the degradation of older social networks based on local production, neighborly support, and barter exchange. It would be accurate to replace statesmen's faces on money with lumps of coal: coal built the modern economy and its ideals of production and consumption. All of these fossil-fueled shifts ripple into revolutions in our energy ontology.

Figure 5.1 Oak Creek Power Plant on Lake Michigan, Wisconsin, USA, 2016. Image by the author.

I Workhouse workhorse

The revolution of steam-driven industry began with stationary engines that pumped water out of coal mines to allow deeper and more continuous operation. Out of necessity, coal had been burned with some regularity for hundreds of years before it was brought into the mine and factory in the 1700s, but this coal was generally gathered from easy-access surface deposits and along the coast, where the ocean eroded cliff faces and coal was jetsam gathered like seashells.

Once steam engines were deployed to drain the soupy depths of mines, a self-generating, positive feedback loop was in place: the engines drained the mines to open deeper coal deposits, which fed the engines that drained the mines that opened the deposits, which fed the engines in the factories, and eventually fueled locomotives and steamships that accelerated global trade. Rather than a fully determinist argument that would elevate the technology itself, the steam engine, as a historical actor that caused specific changes, this self-generating coal-driven economy is a blend of technology and historical context. The British were not dragged into an industrial era with their cravats snagged on a locomotive;

their society made a series of choices that favored innovation by fostering competition and rewarding productivity (Nye 3). John Henry aside, the muscles of an organic economy could not compete with the pistons of a fossil-fueled one. There was nothing inevitable about industrialization in the nineteenth century, but once the wheels set down on an even track, the momentum carried its own self-justifying tone.

Despite the Roman and Victorian vogues of wearing coal as jewelry, "jet jewelry," coal has always brought with it a clutch of negative connotations, most popularly Santa's judgment of bad children. Before the Industrial Revolution that coal made possible, it was perceived as a noisome annoyance associated with disease, death, and the devil (Freese 27–8). Anthracite, a hard form of coal, shares its Greek root with anthrax because the latter is a bacterial disease that consumes the skin in boils that resemble burning coal. Coal itself resembled the buboes, the swollen lymph nodes that wracked European bodies during the bubonic plague of the fourteenth century.

Coal smoke is generally perceived as noxious in odor and color, and long-standing miasmic theories of disease associated its smoke with sinister health effects. In 1257, Queen Eleanor of England complained of the lung-searing coal smoke surrounding Nottingham Castle. When it was used to cook, coal tainted the taste of food and beer (Brimblecombe 8–9). By the time of Queen Elizabeth I, there was a clear lower-class association with coal that her successor King James I sought to dispel by introducing low-sulfur anthracite coal from mines in his native Scotland (Brimblecombe 30). Anthracite coal burned hotter and cleaner (though only relatively so), making it admissible to higher-class hearths. His advocacy for the use of coal was a success, though its effects demonstrated the divergent paths of economy and ecology that endure to this day. By the eighteenth century, coal smoke was inherent to the ambience of British cities, and it colored the sun, rain, fog, and snow with its murk.

The first recorded use of coal was during the Han Dynasty in China, between 200 BC and AD 200. In Belgium, coal was mined in the High Middle Ages, the early twelfth century. By the early fourteenth century, England exported coal to France, and it had opened up most of its coal fields by the middle of the seventeenth century (Smil 187–8). By the eighteenth century, most western European countries suffered an acute shortage of good timber for fuel wood, naval and civil construction, and charcoal. Some statistics point to a much earlier energy crisis from deforestation in the time of Elizabeth I, when cheaper coal increasingly substituted for timber and charcoal (Cipolla 232–3). The English chronicler Edmund Howes observed in 1631: "There is so great a scarcitie of wood through the whole Kingdom, that not only the Citie of London, all haven-towns [ports] and in very many parts within the land, the inhabitants in generall are constrained to make their fires of sea-coal or pit-coal, even in the chambers of honorable personages, and through necessitie, which is the mother of all arts, they have very late years devised the making of iron, the making of all sorts of glasse and burning of bricke with sea-coal or pit-coal" (qtd. in Cipolla 271).

Coal saved England from economic ruin caused by deforestation, and propelled it to the leading position in the race to industrialize (Eberhart xv). Other factors were important in England's success in industry, such as innovations from steam engineers Thomas Newcomen and James Watt, and slave-based labor that steadily supplied raw cotton for textiles. But no engine can be designed without a fuel source in mind, and no factory can be established without a labor force to yoke. Historians have argued that England's industrial victory was not especially due to its clever innovation, but to the sheer luck of its coal-bearing geology. Robert Allen's deterministic conclusion is that "there was only one route to the twentieth century—and it traversed northern Britain" (275).

With coal-driven steam, England led global civilization into a complex industrial order, the implications of which are difficult to exaggerate, so comprehensive have been their effects on the biosphere. The conversion from biomass fuel to fossil fuel in the eighteenth and nineteenth centuries is the single most important shift in the human history of fuel sourcing. These two slender centuries saw a tenfold gain in total energy usage in England, a tremendous leap in consumption compared to the relative stability of per capita fuel consumption for the previous millennia (Smil 187–8).

Along with the huge gains in consumable energy that came with coal, there was a shift in the conception of the nature of energy. The machinists who designed powerful new engines brought a materialist theory of energy that could convert any fuel source—grass, wood, coal—into a calculated work potential based on calories. No longer was food sustenance that flowed transcendentally through the working body; food became fuel, quantifiable and fungible. And the new machines ate coal, as the cows ate grass, as the worker ate the cow; these equivalencies meant, for example, that the calories in the meat of a half-ton steer were equivalent to about ninety pounds of coal.

This calculation sounds rational to the modern reader: conversions exercise control over disparate elements of a diverse world. But it reflects a signal shift in our perception of nature, from unique elements and diverse creatures to a categorical stockpile of natural resources, which would have seemed weird to denizens of a muscular, biomass world (Johnson 44). The literature of this industrial era reflects the sense of exhilaration and vertigo imparted by coal-fired modernity, on one hand, as well as its reductive thermodynamic calculus on the other. Coal literature both extends and qualifies the modern declensionist narrative of nature from an imaginative construct of sublime mystery to a measured standing reserve of natural resources. The positive feedback loop between technical advance and evolving energy ontology enables reductive technical tallies of resources as well as the philosophical reviews of our shifting cultural ideas of what nature is.

Coal use evolved through three distinct innovation cycles. In the first industrial cycle, c. 1787–1814, coal-fired stationary steam engines powered textile mills and pumps. The second cycle, c. 1843–1869, brought with it mobile steam engines, locomotives and steamships, and advances in the efficiency of iron metallurgy. During the third cycle, the industrial world was electrified: between 1898 and 1924, coal was introduced into commercial electricity generation in London and New York, it charged electrical motors in factory production, and brought the first electric motorcars onto the roads (Smil 240).

By the mid-nineteenth century, before the widespread discovery of petroleum and natural gas, coal had a tremendous impact on every facet of society, especially in western Europe and the northeastern United States, which were early to industrialize. It directed urban settlement patterns (dense and largely unplanned), labor practices (unvarying gray hours in factories, which replaced the work-by-daylight schedule and the seasons of agricultural life), the labor force (lamentably including children and slaves), demand for raw materials (often from slave-worked colonies near the equator), the makeup of the wealthy (factory entrepreneurs succeeded landed aristocrats), the units of exchange (from barter to the cash nexus), the atmosphere (soot, smog, and black rains), the landscape (sprawling industrial cities, depopulated countryside), transportation (the railroad and steamships permitted faster transportation on strict timetables), architecture (slender glass-and-iron structures replaced heavy masonry and wood), and interior design (dark-patterned Victorian wallpaper hid coal soot), to name just a few. The infusion of massive amounts of fossil energy from coal was the catalyst of all this bewildering change, a social and economic revolution often cited as the most comprehensive transformation of human existence in our quarter-million-year history.

Philosophers of the early industrial era could not have fully realized the array of implications of global coal burning, which liberated society from the privations of a muscle economy. Historian E. A. Wrigley points out that most Victorian-era essayists failed to recognize coal as a nonrenewable and potentially climate-altering fuel; instead, essays expounded on the invigorated methods of production and distribution or the changes seizing the laboring classes: "Pandora had no prior experience of the irreversible changes which would result from the opening of the jar. For contemporaries the same was true of the Industrial Revolution. Only the generation of Marx, Toynbee, and Jevons was able to gain an understanding of what had taken place" (51). The revolutionary literati of that era, however, often grew nostalgic as the sweeping changes of industrialization took hold. The British Romantic writers are not easy to categorize, but a common thread is that they were political and social revolutionaries but ecological reactionaries. The Romantics looked at the personal freedoms of the past century's naughty coterie courtiers and saw not liberty, but shackles in their oligarchy of finery and philandering. Virtue dwelled with the commoner, and meritocracy was in the air of the new industrial century.

There was more in the air than a spirit of equality. The other revolution that the Romantics witnessed was the "dirty and dusky" revolution of industry, as Byron put it. Compared to their liberal attitudes toward the French Revolution, the Romantics were much more conservative about the revolution that turned the rustic into the machine hand. By the millions, the English poured into cities and factories, and the consequences to landscape, air, and water have been dire and long-lasting. Political revolution and new conceptions of social equality helped facilitate this industrial paradigm shift, since it was seen not only as acceptable but also as desirable for men of no particular birth to use their natural talents to invent machines, develop products, and start companies. The eighteenth-century coterie of gentlemen, generally an underproductive bunch, was eclipsed by a community of citizens hungry for wealth and recognition. Nine thousand miles of new railway in ten years is not the legacy of gentlemen fox hunting on country estates. The new economy was urban, innovative, competitive, crowded, unregulated, and powered by coal.

The Romantic William Wordsworth, leader of the Lake Poets, was a bit of a turncoat when it came to political revolution, turning Tory as heads began to roll during France's Reign of Terror. But he was no apostate when it came to the Industrial Revolution, because he was never an industrial revolutionary in the first place. His raison d'etre as a poet centered on reclaiming the old ways of peasant life, using the "language really used by men" rather than courtier fanfaronade, and advocating an abiding relationship with nature, antipode to the city.

Wordsworth wrote mostly in blank verse, a capacious narrative framework for epic poems, but he also expanded on his poetry with prose-based accounts. In his recollection of the 1814 poem *The Excursion*, he notes:

> What follows in the discourse of the Wanderer upon the changes he had witnessed in rural life, by the introduction of machinery, is truly described from what I myself saw during my boyhood and early youth, and from what was often told me by persons of this humble calling. Happily, most happily, for these mountains, the mischief was diverted from the banks of their beautiful streams, and transferred to open and flat countries abounding in coal, where the agency of steam was found much more effectual for carrying on those demoralising works. Had it not been for this invention, long before the present time every torrent and river in this district would have had its factory, large and populous in proportion to the power of the water that could there have been commanded. (344)

Not only did coal prevent devastation of the remaining English forests, but here we see that its mass adoption into the "demoralising works" of machine-driven industry saved the mountains and streams of the Lake District from hydropower. The "power of the water," often invoked in the emotional passages of his poems that discuss the rivers of that region—the Derwent and the Severn—would have been commandeered by factory owners had not coal been available to fire Midlands furnaces. With factories would have come people to work them, and Wordsworth's happy valley of Grasmere might have become another crowded factory town like Leeds and Sheffield. The old generation loses everything to industrial progress in the overtly pastoral poem "Michael": land, home, livelihood, son. Wordsworth's sentimental hero is drawn as a poignant symbol of sacrifice, though his piteous inability to adapt to changes borders on bathetic sentiment.

Victorian engineers and scientists like Michael Faraday and W. R. Grove adopted a Romantic trope by figuring the technology of the steam engine as essentially organic, merely another medium for the vital, living forces inherent in the natural world (Ketabgian 121–2). After all, the energy that powers a plant and the energy that powers a steam engine have the same ultimate source: the sun. Their thinking influenced Victorian novelists, especially George Eliot, whose scientific acumen was exceptional. Naturalizing coal and the steam engine was one way to neutralize Romantic protests of the adulteration of organic form by suspect mechanical contrivances. By blending man's mechanical genius with the vast organic powers that animated the earth, industrial mechanisms became living, evolving beings that elevated civilization to, and above, nature's level. The many troubling implications of this paradigm include its logical inverse that, if machines are organic, then living beings must also in a sense be machines that can be understood by reduction and dissection. If so, then factory and field workers can be made to labor unrelentingly like machines, as we saw in *Black Beauty*.

The blending of human invention with nature's evolved forms also implied that men could be as gods, using the vast stores of fossil fuel to defy the basic limitations that gravity and time and raw materials placed on production. By seeing the Industrial Revolution's fuel and machines in a continuum, rather than a rupture with the past, Victorian philosophers dressed their era in the appealing organic, vitalist clothes of nature's processes. The rise of the coal-fired steam engine encouraged a wave of "steampunk" science fiction in which the dazzling agencies of the complex machine enabled superhuman feats like time travel. H. G. Wells's *The Time Machine* is an excellent example of a flight of imagination enabled by contemporary ideas of mechanical force, evolution, and organic change inspired by steam power. Charles Babbage's Analytical Engine was steam-driven and designed to perform computer calculus that would eliminate the risk of human error.

The coal-fired Industrial Revolution brought about other major changes in literature. The contrast between urban and rural settings became more pronounced, and many novelists explored the dynamic between these two spheres, such as Charles Dickens in *Bleak House* and Elizabeth Gaskell in *North and South*. The novel of the urban industrial proletariat was born. Thomas Hardy documented shifts in agrarian living as machines came into industrial capitalist agriculture, and the hedgerows were leveled to accommodate the massive upscaling of mechanical farming.

Charlotte Bronte ends her industrial novel *Shirley* with the entrepreneur's prophecy: "The rough pebbled track shall be an even, firm, broad, black, sooty road, bedded with the cinders from my mill" (480). His prediction is fulfilled within a page: "The other day I passed up the Hollow, which tradition says was once green, and lone, and wild; and there I saw the manufacturer's daydreams embodied in substantial stone and brick and ashes—the cinder-black highway, the cottages, and the cottage gardens; there I saw a mighty mill, and a chimney ambitious as the tower of Babel" (481–2). Prominent in these grim depictions of industrial progress is the pollution of air, soil, and water from coal emissions; it is integral to the aesthetic of the nineteenth-century British novel. Not knowing how to neatly interpret these bewildering changes, and torn between the Luddite's lament and the capitalist's crow, Bronte ends her work with a shoulder shrug: "The story is told. I think I now see the judicious reader putting on his spectacles to look for the moral. It would be an insult to his sagacity to offer directions. I only say, God speed him on the quest!" (482).

Thomas Hardy often set his novels in the preindustrial past, or outside the sphere of urban industry, dropping along the way subtle clues of a hulking threat on the horizon. Hardy's rural accounts show how industry was not confined to factory towns, but insinuated itself into the kingdom's remotest locales. Hardy's farmhand Tess suffers in the transition from grass-powered to coal-powered harvests, as new machine threshers transform what had been quiet handwork into a jittering melee: "The hum of the thresher, which prevented speech, increased to a raving whenever the supply of corn fell short of the regular quantity … The incessant quivering, in which every fibre of her frame participated, had thrown her into a stupefied reverie in which her arms worked on independently of her consciousness" (362). Machine-driven work has a scary reverse alchemy of turning living flesh into heat and base metals: namely, the fire and the iron of the steam engine. Tess and company lose their individuality, their communion, and even their minds, to the smart city engineer's gizmo. The futurist is right in claiming that these machines saved labor, including the arduous, repetitive, body-wearing work of agriculture, which requires frenzies of activity around planting and harvesting time. But the machine introduced a new kind of tedium. Tess's Victorian-era generation would have been among the first to experience the somatic shift from the evolved body's desire for conviviality, variety of work, and freedom of movement, to the modern body's dedication to assisting machines at the greatest possible efficiency and speed, the kind of strategy that Henry Ford perfected in his stream-driven assembly lines as he judged his employees with the creed, "the work, and the work alone, controls us" (Johnson 57).

The penny press riffed on the classical pastoral by adapting the traditional love ballad to industrial times. Instead of the shepherd and his lass under a shade tree, industrial-era ballads used the salacious potential of steam-driven friction to entertain the urban poor. In an anonymous street ballad c. 1830, "The Steam Loom Weaver," an engine driver is called in to attend to his lass, "And soon her loom was put in tune / So well it was supplied with steam" (ll. 31–2). The high-energy, clipping cadence of industrial manufacture lends itself well to doggerel:

Her loom worked well the shuttle flew,
His nickers play'd the tune nick-nack,
Her laith did move with rapid motion,
Her temples, healds, long-lambs and jacks,

Her cloth beam rolled the cloth up tight,
The yarn beam emptied soon its seam,
The young man cried your loom works, light
And quickly then off shot the steam.

(ll. 41–8)

Though factory work was often a combination of stultifying and horrific in a time before regulation, the energy infused into daily life and machine metaphors reinforced the popular rhetoric of social improvement and wealth gained through industry. Educated writers not obliged to work in factories posited dire trends of physical and moral degradation, but the workers themselves were more often diverted by the shouted slogans of opportunity, upward mobility, and emigration. And, as ever before and since, by sex and drugs. Escapism was also common: the Pre-Raphaelite artists, working in a time of the most intensive burning of coal seen in history, chose to explore medieval subjects like the King Arthur, and painted in an early Italian Renaissance style. Their technique relied on layering bright colors over a glossy white background, explicitly rejecting the darker, bitumen-based painting that evoked Victorian England.

New coal mines provided fuel for steam engines and iron forges, which together built the iron skeleton of modern architecture. Though most of the iron resulted in utilitarian structures like bridges and factories, iron's ability to support weight with slender lines and organic shapes resulted in whimsical show buildings like 1851's Crystal Palace and 1889's Eiffel Tower, both built to show off metropolitan modernity in world exhibitions. The Crystal Palace's designer, Joseph Paxton, owed his ingenuity to the structure of an exotic water lily at Kew Gardens, the *Victoria regia*. Paxton placed his infant daughter on one of the floating leaves, and instead of sinking under her weight, the leaf proved strong and buoyant enough to support her on the surface (Christianson 79). Paxton deduced that the geometrical venation on the undersurface of the leaves could be recapitulated as transverse girders in an otherwise glass structure. His fusion of a slender iron skeleton with panes of glass became a triumph of modern industrial engineering and a symbol of imperial dominion. The palace housed botanical and cultural collections from the margins of imperial reach, becoming a microcosm of the world's ecologies and cultures.

Like the buildings in other exhibitions, 1848's extant Palm House at Kew Gardens is an elaborate case originally built and heated by coal, the solar energy stored in three hundred million–year-old tropical jungles. The Palm House's climate was supported by a smoke stack and boilers that were removed to a

Figure 5.2 Carl Blechen, *Das Innere des Palmenhauses*, 1832. Image courtesy of Wikimedia Commons. Public domain.

distance where coal delivery via railway and the sooty smoke of its combustion would not spoil the pristine aura of the glass-and-girder tropics. Victorian architecture exploited an aesthetic paradox in coal: this subterranean, filthy mineral was the energetic foundation of the most whimsical, weight-defying, light-loving innovations in modern design. While sunlight-sipping trees made for heavy, dark buildings supported by oak half-timbers in the Tudor style, coal forged spines of iron and skins of glass that admitted the sky.

Its splendid usefulness did not make coal a sweetheart fuel in the nineteenth century, like petroleum in the twentieth century. Intensification of its use merely concentrated its effects and made them seem mundane. In 1830, England mined four-fifths of the world's supply of coal; in 1848 it manufactured more iron than the rest of the world combined; and between 1842 and 1856, the number of coal mines in England quadrupled (Freese 67, 69). Industrial cities became bowls of black fog that curtailed life expectancy due to lung ailments, rickets, and cancers. Coal required special low oxygen conditions to burn, so its fire was usually hidden in a furnace and failed to shed salubrious light. It was vented through narrower chimneys than wood fires needed, so cleaning those chimneys fell to small children, the chimney sweeps who became a symbol of industrial-era oppression.

Also oppressed were the coal miners, whose livelihood depended on daily toiling in a cramped, dark space where there was a continual threat of death from asphyxiation, crushing, burning, and drowning. The mine walls held ghostly, perfect imprints of ferns, reptiles, and even mammals captured in their death poses in a long-extinct habitat, a glimpse of millions of years past that made no sense to the pre-evolutionary worldview of the miner. High demand for coal and the nasty realities of its extraction created new rifts between classes. Though they provided an essential commodity, colliers were often segregated in their own settlements near mines, and this abject social status nurtured solidarity among miners and some of the earliest organized labor movements in the industrial era (Freese 45–6).

Long before mountaintop mining began filling Appalachian valleys with the spoilage of blown-up ridge ecosystems, tunnel mines filled the land with heaps of slag, or spoil tips, which gradually displaced towns. In one dramatic case in 1966, a spoil tip smothered a school in Aberfan, Wales, killing 144 people, including 116 children. Tunnel coal mines routinely crush and gas workers, while surface coal mines deface ancient mountain ranges and cause downstream contamination by salts and metals. Mining coal causes toxic air pollution linked to respiratory ailments and cancer, while burning coal causes the most greenhouse gas emissions of any fuel source. Burning coal causes climate change; the resultant sulfur dioxide causes acid rain.

There is plenty to lament about the deleterious effects of coal. The damage it does to the people who mine and burn it only underscores how terrifically useful it is; we would not keep it an intimate part of our daily society if there were a ready substitute. Coal is so abundant, so cheap, and so energy rich compared to biomass, that it continues to this day as the most important fuel for electricity and stationary industry in most developed countries. This duplicitous promise of coal, its dual convenience and menace, inspires pitched battles between opposed advocacy groups, capturing the postmodern late-capitalist clash between consumption and conservation. Pro-coal lobbies have based their positions not only on claimed economic advantages of coal, but also on various religious and quasi-scientific justifications: God planted coal in the ground to help humans go forth and prosper; the atmosphere is presently deficient in carbon dioxide, and coal acts as an airborne fertilizer needed by hungry staple crops like corn (Freese 191–2). As hydro-fracking has rapidly made natural gas a better economic and, perhaps, environmental alternative to coal, the coal lobby has lost its logical basis and appeals to mining history and culture.

Coal fueled our fires for centuries before there was any conception that its mass combustion could change the global atmosphere. Instead, its smoke served as an indicator of atmospheric patterns, acting as a kind of air-dye that revealed wind and rain movements, the scattering of light, and chemical cycles through soil, air, and water (Freese 166). The carbon long packaged in underground coal veins, liberated by mining and burning, has shifted the carbon dioxide concentration in the air from a preindustrial 280 parts per million to a current 410+ ppm. Most coal today is burned for electricity generation. Coal releases about one kilogram of carbon dioxide into the atmosphere for every kilowatt hour transferred to the electrical grid, to our toasters and laptops. This is about twice the amount of greenhouse gas per watt of natural gas, about twenty times that of solar panels and geothermal sources, and about fifty times that of biomass and nuclear sources (Moomaw et al.).

With its dominant role in industry and its heavy pollution, coal must be considered the most important fuel in the Anthropocene mindset. As we look back on the literature of coal, we have a greater understanding than the authors did of both its enabling energy and its pernicious effects, local and global. But the writers of coal literature are no longer innocent of its role in history. The affect and aesthetics of coal literature show a nascent awareness of the ontological shift into a modern mentality of consumption and pollution, labor and oppression, and the shrinking of time-space into a fossil-powered vehicle.

II The Big Smoke and Auld Reekie

In preindustrial 1659, diarist John Evelyn observed London enveloped in "such a cloud of sea-coal, as if there be a resemblance of hell upon earth, it is in this

Figure 5.3 Joseph Pennell, *The Coal Mine*, 1916. Image courtesy of the National Gallery of Art. Public domain.

volcano in a foggy day: this pestilent smoak, which corrodes the very yron, and spoils all the moveables, leaving a soot on all things that it lights: and so fatally seizing on the lungs of the inhabitants, that cough and consumption spare no man" (qtd. in Allaby 65). Evelyn blamed the pollution on manufacturers rather than the citizens as he observed the "few funnels and Issues, belonging to only *Brewers, Diers, Lime-burners, Salt* and *Sope-boylers*, and some other private trades, *One* of whose *Spiracles* alone, does manifestly infect the *Aer*, more, than all the chimnies of London put together besides" (qtd. in Ackroyd 49). Evelyn's contemporary Margaret Cavendish poetically imagined an early version of the sulfuric acid discovered in coal smoke. She wrote that coal's volatile salts were formed of sharp atoms that corroded all the matter it touched, from lungs to building façades (Brimblecombe 46–9).

Because of its industrial production and the Thames valley's bowl-like effect, London was destined to live under a pall of coal smoke for centuries to come. It was nicknamed "The Big Smoke." Lord Byron's early nineteenth-century image of London from *Don Juan* captures the aura of coal, in contrast to the bright cerulean skies of the hero's native Spain:

A mighty mass of brick, and smoke, and shipping,
Dirty and dusky, but as wide as eye
Could reach, with here and there a sail just skipping
In sight, then lost amidst the forestry
Of masts; a wilderness of steeples peeping
On tiptoe through their sea-coal canopy;
A huge, dun cupola, like a foolscap crown
On a fool's head,—and there is London Town!

(10.9–16)

St. Paul's besmirched dome, then barely a century old, sits atop Ludgate Hill, brooding over its foul nest at the highest point in the old city. The steeples might still be peeping out of the canopy of coal smog, but the parishioners trapped below were asphyxiated by the smoke. In 1810 Louis Simond likened the falling soot to filthy snowflakes that alight on pedestrians in the street: "You just feel something on your nose, or your cheek,—the finger is applied mechanically, and fixes it into a black patch!" (Adkins and Adkins 96). The long century that followed would bring nearly constant coal pollution to London, culminating in the Great Smog of December 1952 that killed twelve thousand people.

The smoke's oppressiveness may show most clearly in its exceptions. Approaching London by carriage on a warm summer day in 1802, William and Dorothy Wordsworth marveled at the clarity afforded when the chimneys were at rest. Dorothy noted how "The city ... made a most beautiful sight as we crossed Westminster Bridge. The houses were not overhung by their cloud of smoke, and they were spread out endlessly" (Adkins and Adkins 96). William wrote an eponymous sonnet that holds a series of praises for the city's beauty, which "doth like a garment wear / The beauty of the morning ... / All bright and glittering in the smokeless air" (ll. 4–6). For the moment, William even prefers the scene over his beloved wilds in the north of England: "Never did sun more beautifully steep / In his first splendour, valley, rock, or hill; / Ne'er saw I, never felt, a calm so deep!" (ll. 9–11). The Wordsworths find pleasure in the world's largest city and industrial center as a "mighty heart ... lying still," where peace is conflated with death, complicating the light images of the thirteen previous lines (l. 14). To William, often oppressed by the city din, a pause in coal-fired industry inspires inner peace, even if it implies the death of its inhabitants.

However suffocating, the smoky effluent of coal was undeniably atmospheric. In the right frame of mind, it was close, cozy, and colorful, perhaps more suggestive to the imagination than a halcyon day in Cadiz. At the beginning of the twentieth century, T. S. Eliot in "The Love Song of J. Alfred Prufrock" memorably captured the "yellow fog" in the form of a cat:

> The yellow smoke that rubs its muzzle on the window-panes
> Licked its tongue into the corners of the evening,
> Lingered upon the pools that stand in drains,
> Let fall upon its back the soot that falls from chimneys,
> Slipped by the terrace, made a sudden leap,
> And seeing that it was a soft October night,
> Curled once about the house, and fell asleep.
>
> (ll. 15–22)

In addition, and opposition, to this cozy and dozy treatment of smoke as a pet, pollution also indicated modernity, productivity, and employment—preferable to the outworn pleasantries of the clear-aired but torpid pastoral. Always adaptable, some denizens of the smoky nineteenth century indicated fondness for these dense, redolent spaces. Edinburgh was nicknamed "Auld Reekie" in honor of its cozy, smelly smoke.

Charles Lamb's 1818 essay "The Londoner" excuses the city's "deformities," namely, its shoddy "gauds and toys" that prey on the avarice of rich and poor alike, because they represent satisfied human appetites (163). Consumer satisfaction comes from industry, and so the self-aware lover of products must in turn love their by-products: "I love the very smoke of London, because it has been the medium most familiar to my vision … Where has spleen her food but in London—humour, interest, curiosity, suck at her measureless breasts without a possibility of being satiated. Nursed amid her noise, her crowds, her beloved smoke—what have I been doing all my life, if I have not lent out my heart with usury to such scenes?" (163). Lamb's delight is playfully economical, in step with his surroundings. His sympathetic heart gains interest as he loans it out in order to perceive the tussle of these business streets. His occupation is of a heart loaner, preserving and enriching the humanity of the factory city in his writings.

Unlike the common hypocrite who wishes to consume industrial goods without suffering the industrial atmosphere, Lamb embraces the "beloved smoke" and the smoking machines of industry. Charles Dickens lampoons the hypocrisy of wealthy consumers who delight in the money and "elegancies of life" provided by industry but, as with Mrs. Gradgrind in *Hard Times*, "could scarcely bear to hear [Coketown] mentioned" (26). Lamb and Dickens have marked the era when self-involvement in the greater industrial project bifurcated into those who take accountability and those who discard it. They write some of the first fictional critiques of the NIMBY (not in my back yard) hypocrisy of industrial environmentalism.

Lamb's "multitudinous moving picture" included such lugubrious sights as the thousands of child peddlers living in the shadows of city streets. Children made meager careers sweeping street crossings, selling vegetables, and peddling baubles. The rapid shift from rural agrarian to industrial urban life meant there were few social services to protect vulnerable children, many of whom were orphaned due to the high risk and health impacts of factory work and waves of typhoid and cholera that spread virulently in urban slums. Lamb the Londoner wrote proudly, "I was born (as you have heard), bred, and have passed most of my time, in a *crowd*. This has begot in me an entire affection for that way of life, amounting to an almost insurmountable aversion from solitude and rural scenes" (162). Quite unlike his Romantic contemporaries who usually found their subjects among rural people and wild scenes of nature, Lamb celebrated the "mob of happy faces crowding up at the pit door of Drury-Lane Theatre" and found his hypochondria deterred by immersion in the crowds on the Strand, where "I fed my humour, till tears have wetted my cheek for inutterable sympathies with the multitudinous moving picture, which she never fails to present at all hours, like the shifting scenes of a skilful Pantomime" (162–3). Lamb's essay "In Praise of Chimney-Sweepers" demonstrates the empathetic eye with which he viewed the urban children that were part of a new industrial work force. "Chim chimminy Chim chimminy Chim Chim cheroo"; the cheery melancholy of urban England was on its way to mainstream entertainment, first through novelists like Dickens and then in the twentieth-century nostalgia of Disney's Mary Poppins.

Charles Lamb's view of the city is unusually laudatory among Romantic accounts of London in the nineteenth century. Leigh Hunt, Lamb's contemporary, would observe the "yellow atmosphere of money-getting" as a dual ecological and economic symbol of the times (qtd. in Edgecomb 29). In 1846, the American poet and traveler J. Bayard Taylor published *Views Afoot*, which details the unique and surreal qualities of industrial London's weather. The gloom frustrates any aesthetic study, since "no light and scarcely any warmth can penetrate the dull, yellowish-gray mist, which incessantly hangs over the city" (306). The occasional "sickly gleam of sunshine" is "soon swallowed up by the smoke and drizzling fog. The people carry umbrellas at all times, for the rain seems to drop spontaneously out of the very air, without waiting for the usual preparation of a gathering cloud" (306). While these conditions are depressing and disgusting, they also stimulate the morbid imagination in a way that Charles Dickens and

Arthur Conan Doyle would come to appreciate. According to Taylor, November in London "resembles the gloom of Hades. The streets were wrapped in a veil of dense mist, of a dirty yellow color, as if the air had suddenly grown thick and mouldy. The houses on the opposite sides of the street were invisible, and the gas lamps, lighted in the shops, burned with a white and ghastly flame" (306). The general darkness pierced by an unnatural gas-fueled light reminds Taylor of "Satan resting in the middle of Chaos" (383). Taylor's narrative of The Big Smoke was a familiar one among foreign travel writers visiting the industrial capital of the world. Nineteenth-century London was cloaked in coal smoke, but also bejeweled with the products of its success: the dizzying plethora of consumer goods. Coal brought to the world an unprecedented collision between economic success and eco-social oppression, a tension that tended to resolve itself by further social stratification.

Friedrich Engels published his classic *The Condition of the Working Class in England in 1844* just a year after recording his observations, but it was not translated into English for half a century because of its searing indictment of industrial labor. Engels lends an outsider's view of England where the initial impressiveness of the scenes, "hundreds of steamships dart rapidly to and fro," is merely the bustling surface of the deeper reality of "the human suffering which has made all this possible. [The visitor] can only realize the price that has been paid for all this magnificence after he has tramped the pavements of the main streets of London for some days and has tired himself out by jostling his way through the crowds and by dodging the endless stream of coaches and carts which fills the streets ... The restless and noisy activity of the crowded streets is highly distasteful, and it is surely abhorrent to human nature itself" (64). Rather than Lamb's engaged sympathy and affinity, Engels sees a flow of automatons who "rush past each other as if they had nothing in common"—an observation close to DeQuincey's distaste for railway stations (64). This crowd rush is driven by coal and its steam power, which obscures the natural vicissitudes of day and night, summer and winter, with the staring clock face and its inexorable hands.

Engels's point of engagement with the "whirlpool of modern industrial life" is the suffering of the working class, whose "wages are so low that they can hardly keep body and soul together" (58). Manchester is "the most important factory town in the world" because it clearly exhibits, on one hand, "the degradation into which the worker sinks owing to the introduction of steam power, machinery, and the division of labor" and, on the other hand, "the strenuous efforts of the proletariat to raise themselves from their degraded situation" (57). The city is built on coal, "blackened by soot," and filled with the "helots of modern civilization"—the industrial serfs striving to improve their condition in an unjust social order (79). Though Engels is generally focused on people, a contemporary reader can also identify the manifold negative environmental impacts of coal-driven industry in his description of the river Irk: "At the bottom the Irk flows, or rather, stagnates. It is a narrow, coal-black, stinking river full of filth and rubbish which it deposits on the more low-lying right bank. In dry weather this bank presents the spectacle of a series of the most revolting blackish-green puddles of slime from the depths of which bubbles of miasmatic gasses constantly rise and create a stench which is unbearable even to those standing on the bridge forth or fifty feet above the level of the water" (78). The "rookeries" of the poor teeter at the cliff edge of the river in the shadow of the workhouse, which is "built on a hill and from behind its high walls and battlements seems to threaten the whole adjacent working-class quarter like a fortress" (80). This formative experience in Manchester directed Engels's radical advocacy of workers' rights, and the very next year, 1844, Engels met Karl Marx and they began their triumphal collaboration on behalf of the proletariat.

North and South, Elizabeth Gaskell's novel of the industrial Midlands, records the economic treadmill of factory production enabled by an easy supply of coal fuel. "At Milton the chimneys smoked, the ceaseless roar and mighty beat, and dizzying whirl of machinery, struggled and strove perpetually. Senseless and purposeless were wood and iron and steam in their endless labours" (418). Mimicking the tireless machinery, the workers become robots to keep their positions despite constant threats of replacement "but the persistence of their monotonous work was rivalled in tireless endurance by the strong crowds, who, with sense and with purpose, were busy and restless in seeking after—What?" (418). Gaskell captures the sense of modern capitalist ennui setting in, where time is reduced to a changeless gray limbo, and enterprise drones through seasons and years without ever arriving at a promised goal. The only goal is to produce more, more cheaply, and not to be outcompeted. It ends only when a company goes out of business.

Unique to the industrial period, this feeling of overstimulation, exhaustion, and monotony was made possible by the continual supply of fuel to the factories. Before fossil fuels, biomass required seasonal vicissitudes where cycles of growth, gathering, processing, and burning offered a modicum of change. Seasonal variation was preserved in sowing and reaping, and years were distinguishable from one another based on the fortunes of the harvest. To an industrial worker, these changes were experienced only as price variations—the cost of a loaf of bread based on the wheat harvest—and the consequent struggle of feeding a family on a fixed income. In a textile mill, the mainstay of the British industry, the factory work itself did not change from summer to winter, beyond sweating or shivering. There was either more of the same work, or destitution.

Victorian writers repeatedly turned to toxic urban fogs for emotion-inflected settings. This trend indicates an increasingly acute concern among writers in the second half of the nineteenth century. The trope of apocalypse, still fashionable in the twenty-first century, is the gesture of the powerless writer to "imagineer" the overthrow of a corrupt world order. When Richard Jefferies imagined industrial society literally sinking into its own filth in *After London*, his postapocalypse hero Felix learned of the hazards of the lingering toxic fog from the flight patterns of instinctual birds. These toxic fog works suggest that the greed and vice of getting and spending was closely related to the pollution of Victorian England. Nature actively corrected this industrial egotism, with a vengeance. The same desire to overthrow the corrupt world order operates in today's eco-apocalypse genre. Eco-apocalypse writing imagines biosphere-level alteration of climate to be the driving force of global housecleaning, rather than localized pollution from industry.

The Victorian essayist John Ruskin was a perceptive critic of the arts. His observations of the bewildering changes reflected in art, such as the smoggy aura of J. M. W. Turner's paintings, led Ruskin to make social and moral commentary on his age. His definition of merit in art is that the greatest artist "has embodied, in the sum of his works, the greatest number of the greatest ideas" (*Stones of Venice* 92). Far from the fancies of the pure aesthete, Ruskin's favorite pieces are social, political, and historical. Instinctively nostalgic, Ruskin continually turned to the values of slow time and the celebration of the soul inherent in preindustrial handcraft, endangered in the nineteenth century by anonymous factory mass production. His concerns extended from the fate of the craftsman's soul to the conversion of nature into a technological medium for modern production: "How much of it do you intend within the next fifty years to be coal-pit, brick-field, or quarry?" (*Works* Vol. 3 337). Imagining their success, Ruskin draws an image: "From shore to shore the whole of the island is to be set as thick with chimneys as the masts stand in the docks of Liverpool: that there shall be no meadows in it; no trees; no gardens; only a little corn crop grown upon the housetops, reaped and threshed by steam" (*Works* Vol. 3 337). In our contemporary view, we might assume Ruskin is concerned about habitat degradation and species extinction. However, Ruskin is principally worried that "beautiful art can only be produced by people who have beautiful things around them" (*Works* Vol. 3 338). Instead of pastoral beauty, Ruskin sees "the furnaces of the city foaming forth perpetual plague of sulphurous darkness; the volumes of their storm clouds coiling low over a waste of grassless fields, fenced from each other, not by hedges, but by slabs of square stone, like gravestones, riveted together with iron" (*Works* Vol. 3 339).

Beautiful art did come out of this coal squalor. Turner, Ruskin's favorite artist, turned from pastoral landscapes to industrial ones later in his career in paintings like "Rain, Steam, and Speed." Monet's richly smudgy views of industrial London are among his most admired works. Without urban pollution, what would Dickens have written about, and where would it have taken place? Coal smoke permeates his work from London to Coketown. Though it has its moments of menace, especially works like *Hard Times*, coal smoke is most often something that envelops his characters in an ambience of ambiguity, mystery, sympathy, and perseverance. The working classes would have no work and no story without coal smoke. In *Bleak House,* the atmosphere is onerous, but Dickens's signature humor makes the scene almost cheery: "Cook's Court was in a manner revolutionized by the new inscription in fresh paint, PEFFER AND SNAGSBY, displacing the time-honoured and not easily to be deciphered legend PEFFER only. For smoke, which is the London ivy, had so wreathed itself round Peffer's name and clung to his dwelling-place that the affectionate parasite quite overpowered the parent tree" (154). Parasitical, perhaps, but in a loving way smoke adorns buildings and gives the city character. Like Charles Lamb, Dickens perceives that coal smoke is the urban aesthetic counterpart to the natural ivy that adorns the structures of Oxford and Cambridge, and the city worker, lacking university education, may still find evidence of his work by observing coal smoke clinging to the walls. Smoke is industrial society's baby, the grandchild of ingenuity, ambition, raw material, and fossil fuel.

Though coal is most famous for blackening The Big Smoke and Auld Reekie, it also produced the opposite effect: bright light. Deriving coke from coal allowed the black rock to replace charcoal in blast furnaces, thus forestalling further catastrophic deforestation. This invention had a second virtue: a by-product called coal gas was useful for illumination. Coal gas, a synthetic form of natural gas, was a Victorian segue fuel used between the long era of darkness, window candles, and oil lamps in city streets, and the modern era of electricity that came to major cities by the start of the twentieth century. Gas street lamps first appeared in 1812 in London, and the last oil lamp in modish Grosvenor Square was replaced by gas in 1846 (Picard 66). The reign of coal gas did not last long: the first electric lamps in London appeared on the Embankment in 1878 (Nead 83).

Gaslight appeared ghostly. Its likely etymology dates back to the Dutch *geest*, meaning spirit (Gordon 20). At first a source of bedazzlement, coal gas lamps within a few decades came to seem lurid and filthy compared to the newer alternative in electricity. Still, this brief reign of gas illumination made an imprint on literature. Its concentrated use during the Victorian era provided a unique ambience that can be described in artistic terms as a living chiaroscuro: spots of blanching light surrounded by deep pools of shadow. Aesthetically, we could compare the Victorian experience to our current distinctions between harsh fluorescent and warm incandescent light bulbs. Newer LED technology is greatly invested in the aesthetic quality of its light, continually innovating toward wide dispersal and warm tones.

Gaslight was bright and its resource was abundant, but those virtues were also liabilities in lived experience. Like the exchange of open wood hearths for closed iron coal stoves in the domestic space, the rise of coal gas lamps brought to the streets a luminary shift that directly translated into an emotional one. As with the steam-driven technologies enabled by coal, the coal–gas visual aesthetic is a feedback loop between the states of two Victorian arts: the technology of fuel combustion and the philosophy (ontology and aesthetics) of writers living in the glare. The warmer, dimmer tones of candles and oil lamps were overwhelmed by the interrogatory glare of the white gaslight. As Romantic revolutionary literature succumbed to the straightened, proud morality of the Victorian era, the Romantic reign of candlelight and whale oil died of exposure to Victorian coal gas.

In the Victorian literature of coal gas, the very fuel that was to bring safety and orientation to the night streets ironically became associated with secrecy, evil, and irrationality (Nead 83). For example, the first glimpse of Stevenson's diabolical Mr. Hyde comes by gas lamp, when he tramples a small child in the midnight streets of London. The witness describes walking at "about three o-clock of a black winter morning, and my way lay through a part of town where there was literally nothing to be seen but lamps. Street after street, and all the folks asleep—street after street, all lighted up as if for a procession and all as empty as a church—till at last I got into that state of mind when a man listens and listens and begins to long for the sight of a policeman" (6–7). The character Enfield is experiencing the uncanny in his 3 a.m. ramble: that crossroads between mind and space in which repetition leads to disorientation, as the physical body's displacement blurs the mental boundaries between the dream and waking world. Once confounded in dream, perceptions take on the symbolic and monstrous qualities of the supernatural, so a grid of well-lit streets transmutes into a maze in which brightness only excites confusion.

No cop appears to reorient him; instead that strange chimera Hyde further perturbs his perception by stalking onto the scene and committing an act of coded violence against a child. Enfield's story casts a spell over the lawyer, Utterson, who becomes the main investigator of Hyde's crimes. That night Utterson has a nightmare featuring a cinematic "scroll of lighted pictures" that swirls toward vertigo as the dreamer moves "the more swiftly and still the more swiftly, even to dizziness, through wider labyrinths of lamplighted city, and at every street-corner crush a child and leave her screaming. And still the figure had no face by which he might know it; even in his dreams, it had no face" (39). The lamps obliterate the perpetrator Hyde's face with garish light.

A leitmotif of cold, unnatural illumination continues throughout the story: wind perturbs the light and shadow patterns in a way that seems to bring alive malfeasant energies in the city. The lamps "glimmer like carbuncles" through a fog that seems to drown London's pedestrians (52). Soho's "lamps, which had never been extinguished or had been kindled afresh to combat this mournful re-invasion of darkness, seemed, in the lawyer's eyes, like a district of some city in a nightmare" (48). When Hyde comes to the point of confession, his recollection of the joy of murder is again dressed in the invasive light of the lamp: "A mist dispersed; I saw my life to be forfeit; and fled from the scene of these excesses, at once glorying and trembling, my lust of evil gratified and stimulated, my love of life screwed to the topmost peg. I ran to the house in Soho, and (to make assurance doubly sure) destroyed my papers; thence I set out through the lamplit streets, in the same divided ecstasy of mind, gloating on my crime, light-headedly devising others in the future, and yet still hastening and still hearkening in my wake for the steps of the avenger" (87). Hyde/Jekyll is here both pursuer and pursued, and the lamps illuminate the wide-ranging city and make it available to a creature of the night who stalks an alternative, nightmarish territory totally foreign to its sunlit counterpart. The story does more than just show an antagonism between night and day; Stevenson is clearly using his firsthand experience of the modern gothic of coal-lit streets to elaborate an uncanny, violent environment enabled by the rise of the gasworks.

The novel's tensions between good and evil, overtness and secrecy, revelation and ignorance continually play off the actual spectacle of white-black gaslight. Rather than extirpate violence from city streets, gaslight seems to illuminate a stage for immoral behavior and threatening surreality. Gaslighting was associated with civic improvement because it extended hours of trade and permitted extravaganzas like the Cremorne Pleasure Gardens of the 1860s, when the nighttime city was laid bare under nearly half a million streetlights.

But it also had the effect of revealing, in lurid detail, the cloak of coal smoke on buildings inside and out, the flaws in beloved faces, the sleights of hand among shopkeepers, gristle in meat and bruises on the fruit of street markets (Naed 88). These paradoxes embody the visual complexities of coal-fueled cities, where the atmosphere is deepened into literal layers of pollution, figurative layers of morality, and the subjective feud between ugly and alluring.

III In depth: *Wuthering Heights* and *How Green Was My Valley*

Urban landscapes were most clearly marked by coal and most often explored by writers, but the economy and ecology of coal-producing regions also merited literary study. That dark and dire Victorian classic, *Wuthering Heights*, has a subtle vein of coal running through the text. Emily Bronte's novel takes place in Yorkshire, England's largest county, known in her time for its productive coal mines. It was also close to rising industrial cities like Sheffield, Leeds, and Rotherham, the locales that led the world in textile manufacturing. Bronte knew of what she wrote: her home in the vicinity of Haworth, in west Yorkshire, included a remote farmhouse named Top Withens, which was thought to be the inspiration for the brooding bulk that housed Heathcliff.

   Wuthering Heights and its environs are a clear metonym for their master. The Romantic hero Heathcliff, with his blend of smoldering intensity and chilling callousness, stands in stark contrast to his refined rival, Edgar Linton, who lives at the contrasting abode, Thrushcross Grange. (A Monty Python sketch uses semaphore flag language, the kind used by beach lifeguards, to capture the dualism in absurdity: characters communicate from Heights to Grange with emphatic flag waves.) Bronte makes the land-to-man contrast overt when Catherine is courted by Edgar Linton: "Doubtless Catherine marked the difference between her friends, as one came in and the other went out. The contrast resembled what you see in exchanging a bleak, hilly, coal country for a beautiful fertile valley; and his voice and greeting were as opposite as aspect. He had a sweet, low manner of speaking, and pronounced his words as you do: that's less gruff than we talk here, and softer" (70). Heathcliff's landscape name directly evokes his character: bleak, hilly, coaly. Linton's name is not analogously pastoral, but it echoes "lintel," the decorative horizontal block that spans the gap in a doorway and a fireplace. It is an essential structural element that evokes entrance, support, shelter, and hearth fire.

Nelly's characterization of Heathcliff as coal country and Edgar as a fertile valley comes from Catherine's own sense of the two men. Catherine tells Nelly: "My love for Linton is like the foliage in the woods: time will change it, I'm well aware, as winter changes the trees. My love for Heathcliff resembles the eternal rocks beneath: a source of little visible delight, but necessary. Nelly, I *am* Heathcliff!" (82). Her deciduous love for Edgar is a season's passing, leaves falling without any promise of a returning spring, but Heathcliff is the bedrock and, in Yorkshire, the coal contained in the "eternal rocks beneath." Like the fuel, Heathcliff excites little delight but is essential to the life of this coal country. Heathcliff is racially other, described as dark and gypsy (though he may be Earnshaw's bastard son); he is cast into a role of toil and dirt by the family; slowly, inexorably, he gains control over the economy of the Grange and the Heights by sheer determination. Heathcliff is cold coal, sparked to flame by Catherine and, suffering her loss, he smolders down to embers and burns everyone who encounters him.

At Wuthering Heights, the hottest coal fires cannot warm the ambience of the troubled domestic space. At the beginning a curious Mr. Lockwood, tenant of Thrushcross Grange, is repelled from his nest by an inept "servant-girl on her knees surrounded by brushes and coal-scuttles, and raising an infernal dust as she extinguished the flames with heaps of cinders. This spectacle drove me back immediately" (9). This humble scene captures the daily inconvenience of living with coal as a domestic fuel, even in a pleasant valley house managed by servants. Lockwood is the fortunate one who has a choice to retreat from the choking dust, and he sets out for the clear-aired heath and Wuthering Heights.

Chilled by his journey, Lockwood seeks shelter under his landlord's roof and, after being sauced about by Joseph and Hareton, enters the main living room: "We at length arrived in the huge, warm cheerful apartment, where I was formerly received. It glowed delightfully in the radiance of an immense fire, compounded of coal, peat, and wood; and near the table, laid for a plentiful evening meal, I was pleased to observe the 'missis,' an individual whose existence I had never previously suspected. I bowed and waited, thinking she would bid me to take a seat. She looked at me, leaning back in her chair, and remained motionless and mute" (10). Despite the seeming welcome of light and heat from the multi-fueled fire, Lockwood finds no warmth in the company. Young Cathy's hearth burns like an inferno rather than a domestic scene of refuge from an outside snowstorm. Mistaking her for Heathcliff's wife rather than a despised embodiment of the union between Edgar Linton and Catherine, Lockwood mistakes the relationships in the hilltop house just as he misreads the "warm cheerful apartment" that "glowed delightfully in the radiance of an immense fire." His misses are soon corrected by her blank coldness. Not even an inferno driven by three distinct fuels can melt the chilly presence of Heathcliff, the antihero with a penchant for hanging puppies.

At Wuthering Heights, hearth fire is hellfire. After spending a night huddled in the kitchen, Lockwood enters into this inner chamber once again, "where the females were already astir. Zillah urged flakes of flame up the chimney with a colossal bellows; and Mrs. Heathcliff, kneeling on the hearth, reading a book by the aid of the blaze. She held her hand interposed between the furnace-heat and her eyes, and seemed absorbed in her occupation; desisting from it only to chide the servant for covering her with sparks, or to push away a dog, now and then, that snoozled its nose over-forwardly into her face. I was surprised to see Heathcliff there also. He stood by the fire, his back towards me, just finishing a stormy scene to poor Zillah" (30). Though everyone, even the dogs, have gathered around the fire for its warmth, it annoys with excessive heat and sparks. The fire is scarcely differentiated from Heathcliff himself, who spews his animosity with the blaze as his background.

Later, Isabella Linton, the immiserated wife of Heathcliff, remarks on her first visit to her new home: "There was a great fire, and that was all the light in the huge apartment, whose floor had grown a uniform grey" (138). Soon the master's stationing repeats itself when Heathcliff "took his stand on the hearthstone, near me, and began to put questions concerning Catherine" (146). The master has no qualms about degrading his new wife at the expense of the woman he truly loves, and his open dictates of disavowal are made more forceful by his position "on the hearthstone," the symbolic heart and literal power station of the house. Time passes, and the rage abates as his change toward a final happy imbecility approaches. He talks to Nelly of the dead and haunting Catherine: "His eyes were fixed on the red embers of the fire, the brows not contracted, but raised next to the temples; diminishing the grim aspect of his countenance, but imparting a peculiar look of trouble, and a painful appearance of mental tension towards one absorbing subject" (291). His anger, passion, torment, and obsession are united in embers of the coal fire.

These hellfires are Heathcliff's animus. Another, more leisurely time, the narrator Nelly encounters the servant Joseph, who "seemed sitting in a sort of Elysium alone, beside a roaring fire; a quart of ale on the table near him, bristling with large pieces of toasted oat-cake; and his black, short pipe in his mouth" (236). Joseph takes unkindly to being interrupted from his moment of peace amid the general violence of Wuthering Heights—his Elysium recalls the pleasure of hearth fires under normal conditions. Soon the violent fires in the hearth will abate with Heathcliff's death, and Wuthering Heights will return to a pleasantly ghost-free space where innocent young love may kindle.

As Lockwood approaches the Heights in this final scene, "both doors and lattices were open; and yet, as is usually the case in a coal district, a fine, red fire illumined the chimney: the comfort which the eye derives from it renders the extra heat endurable. But the house of Wuthering Heights is so large, that the inmates have plenty of space for withdrawing out of its influence; and accordingly, what inmates there were had stationed themselves not far from one of the windows" (307). At last, with the death of Heathcliff and his interment next to Catherine, their bones rest amid the fossil coal seams of their homeland. The living left behind are released from the spell that was cast with Heathcliff's arrival, and these home fires are not the old diabolic infernos, but a quality of "fine, red fire" that gives comfort even on a summer night. Rather than jailing inmates clustered around a too-hot fire, Wuthering Heights now expands into a continuum with its pleasant surroundings, windows open, issuing welcome to approaching strangers.

Heathcliff may be a coal fire, but a coal fire is not always Heathcliff. By the end of the novel, the harsh landscape of Wuthering Heights is resolved and softened by its loving tenants, so the coal itself can transmute from a holocaust to a comforting red fire well-aired by the capacious house. The stark dualisms that drive the novel, headed by the opposing abodes of Heights versus Grange, drain away into a gentle continuum between house and landscape, the kindling of comforting coal. It is a symbol that unifies Bronte's entire novel and serves as a leitmotif of the nineteenth century: how to transform diabolical energy into a fuel that is salubrious and productive, a creative rather than destructive force. Coal is inherently both.

In figuring Heathcliff through a metaphor of fuel, Bronte deploys the natural resources of her native Yorkshire by invoking them as human energy. This figurative use of "energy" became popular in marketing in the twentieth century. Human energy, say the petroleum companies, is mighty enough to supersede our challenges with the limited supply and pollution of *actual* energy, fuels like petroleum. Bronte's metaphor works in the opposite rhetorical direction, convincing readers that Heathcliff is diabolical by using the vehicle of coal—too hot, smoky, sparky, and ash-dusty. Heathcliff's energy might burn the Heights to the ground. It's not a vague promise of a bright, innovative future, but an embodiment of a devilish present. Jane Austen used wood fires to characterize the generosity of heroes like Bingley, but she never implied that Bingley *was* a warm, bright wood fire. Here, half a century later at the height of industrial expansion in England, we see Bronte taking that step of imagination in which people *become* fuel. This is a much more radical assertion than simply placing characters within historical and social conditions; it asserts a changed ontology of the Anthropocene. We *are* the fuel that we consume—our brains and bodies are continuous with the machines and infrastructure that characterize a world of fossil fuels. Coal's energy and our energy are indistinguishable.

By the early twentieth century, coal mines had opened nearly every available seam in Britain, and the industry grew in America's Appalachian valleys along with increasing class disparities between rural miners and urban consumers. D. H. Lawrence's father was a coal miner, and his book *Sons and Lovers* may indicate his alienation from his father and lament for his intellectual mother's lost opportunities. Towns existed only to produce coal for the industrial and domestic markets, like Richard Llewellyn's Welsh valley town in his 1939 novel *How Green Was My Valley*, which takes place mostly in the 1890s. Llewellyn's account of the lives of miners came largely from his residence at Gilfach Goch ("Red Nook"), a mining village in South Wales nestled in a valley between pastoral hills. His narrator Huw Morgan's *bildungsroman* is shadowed by the mine's slag heap, the pile of waste rock and shale excavated by mining. Though on the surface the story involves Morgan's development, exhibiting the passions, pleasures, and struggles of a stripling, a darker vein runs through the tale, showing the trials of the workers and the mine's gradual despoilment of the valley.

Spoken in a lyrical Welsh dialect, the novel gives a traditional patriarchal account of Welsh mining life. The hardworking men see their livelihoods, village society, and subsistence in nature collapse as the coal industry chugs toward the twentieth century. The novel never leaves this Welsh valley, except by imagination. London merits frequent mention, two of the brothers emigrate to America, the sons sing for the ancient Queen Victoria, and one of the Morgan sisters moves with her husband to South Africa during the Boer Wars. But Huw, the most scholarly son in the family, by choice stays to work in the mines and inherit the kingdom of slag and ruin.

There is no mention of the slag heap until a hundred pages into the five-hundred–page novel, when Huw steps out of his retrospect into the present tense:

> The slag heap is moving again.
>   I can hear it whispering to itself, and as it whispers, the walls of this brave
> little house are girding themselves to withstand the assault.... for those months

the great bully has been beaten, for in my father's day men built well for they were craftsmen. Stout beams, honest blocks, good work, and love for the job, all that is in this house.

But the slag heap moves, pressing on, down and down … Soon, perhaps in an hour, the house will be buried, and the slag heap will stretch from the top of the mountain right down to the river in the Valley. Poor river, how beautiful you were, how gay your song, how clear your green waters, how you enjoyed your play among the sleepy rocks. (102)

Like generations of Victorian writers before him, Llewellyn plays the pastoral scene against its aesthetic antipode, the industrial wasteland carved out by coal production. His images prefigure the spoil heaps tumbled into river valleys by modern mountaintop removal practiced in Appalachia. The slag heap rears itself over the story slowly, and is mentioned only sporadically. But in the end, it topples and suffocates Huw's world. The crushed craftsman's cottage is a symbol of the dedicated work of former generations literally interred in coal spoil. Their lives are crushed by their livelihood. It is a scene of primary pollution—the impacts of mining at the source—rather than the secondary air pollution that comes from coal burning. As a personal account, the novel testifies about the often-hidden effects of mining practiced in small communities far removed from the population centers that provide the coal market. The novel engages no discussion of the reason coal is being mined, where it goes, who consumes it, for what purposes; the only concern is the nature/culture of this coal nook.

Llewellyn develops the dialectic between nature and spoil through time. As the slag heap grows, nature adorns the waste with its flowers and grasses, but the pace of human production far outstrips nature's restoration:

That way [by the river], I had to pass the two heaps of slag that had grown and grown till they looked half as big as the mountain. Even grass was growing in some places, as though to take pity on us and cover the ugliness of them. The river running between was drying up, so sick it was of the struggle to keep clean, and small blame on it.

Farther on, past the last of the cottages, green grass grew again, and happy it was to see a flower growing after all the brute sadness, though the river was still running black and the plants and reeds dead and dying on both banks. (171–2)

The engines pipe bilge directly into the river, and the slag pile that started by the mine entrance slugs its way uphill through the village streets. Solid and liquid pollution are killing the ecosystem, which had existed in harmony with the village gardens and fishing spots. The valley is becoming uninhabitable for everything but the mine spoil, so Huw calls it a "bully" that pushes more delicate forms out of its heavy way.

Huw's disgust for the waste of coal mining and the harshness of the work itself does not sow hatred for the miners or the valley they live in; in fact, Huw's love deepens as he finds solidarity with the other workers in the middle of hardship—a solidarity so palpable in present-day coal mining communities. Though he excels in school and has the opportunity to go into white-collar work, he stolidly chooses the mineshaft with his brother and father. There he learns the toil:

> For hour after sweating hour, bent double, standing straight only when we were flat on our backs, we worked down there, with the dust of coal settling on us with a light touch that you could feel, as though the coal was putting fingers on you to warn you that he was only feeling you, now, but he would have you down there, underneath him, one day soon when you were looking the other way.... I always thought I saw a face in the glitter of the coal face, and never mind how much Ivor cut from it, it always seemed to be there. (373)

Coal miners' bodies have to adapt to an array of weird, unevolved conditions by physical repression of the natural needs for light, air, and space (Johnson 64). In Huw's charmed imagination, he has conversations with inanimate things: the cottage, the river, the slag heap, the face of coal itself. He is haunted by destiny, clearly perceiving that his fate is to be crushed by coal or slag—he seems to acquiesce. His brothers (other than Ivor) fled from the valley to seek other lives, ones that consumed coal rather than produced it. But Huw, his brother Ivor, and his father Gwillym stay at the pick. As their picks do violence to the coal seam, the face in the coal bears witness to their debt. The glittering black face mimes: death is near. The anxiety and melancholy of this mind-space is typical of miners. Their bodies have already become denizens of a coal world, with blackened skin, dusty lungs, and the mole's motions. They are the ancestors of Morlocks, H. G. Wells's underworld post-humans whose adaptations have made the bright, airy surface world sensually intolerable.

With his family thinned out by emigration and fatality, a final ominous move comes from the slag heap. With no room left in the valley for spoil, the mine company sends it by conveyor belt to the top of the mountain where it piles around the rim of the valley. To Huw's protest, the worker explains, " 'If you want to work, the slag must come out. If it comes out, it must have a place to go. So there, you, and here it is.' " Ever gentle-minded, Huw thinks, "No use to blame him," and returns to his family cottage cowering beneath the waste mountain.

The novel concludes with a conflation of wild nature and the man-made mine, which had been opposed:

> There is patience in the Earth to allow us to go into her, and dig, and hurt with tunnels and shafts, and if we put back the flesh we have torn from her and so make good what we have weakened, she is content to let us bleed her. But when we take, and leave her weak where we have taken, she has a soreness, and an anger that we should be so cruel to her and so thoughtless to her comfort. So she waits for us, and finding us, bears down, and bearing down, makes us a part of her, flesh of her flesh, with our clay in place of the clay we thoughtlessly have shovelled away. (497)

Llewellyn depicts nature as an injured mother taking revenge for her violation. The slag is her hammer coming down in verdict. This active characterization is a shift from the narrator's earlier views of despoilment, where nature was passive to man's activities and Huw's shame for the pollution and his pity for nature were just liquid emotions easily diverted by the mine machine. In the end, Huw the old man feels culpable for the work that ruined the valley; with satisfaction he imagines nature's crushing reprisal. At least it is a victory for something living, rather than an apotheosis of waste.

How Green Was My Valley shows a struggle between mine and nature, worker and owner, which was prescient of future losses. Llewellyn's fiction of the crushing slag heap came true in a 1966 disaster in Aberfan, Wales, when a mine spoil tip slid into the valley town and killed 144 people, 116 of them children in the local school. In this case, the waste pile on the mountain slope was built on top of natural springs that fed into a heavy slurry that surged downhill. The second most deadly landslide from coal waste occurred in Buffalo Creek Hollow, West Virginia in 1972. Half a million cubic meters of black waste slurry broke through an impoundment dam and surged downhill, suffocating everything in its path. The disaster killed 125 people and left several thousand residents homeless.

The literature of coal has become a trope that reflects social inequality based on unequal access to the goods and revenues coal energy made possible. The most enduring symbol of the promise and trauma of modernity, coal narratives evoke the dualism of imprisonment and freedom held within the chemical bonds of this material afterlife of ancient sunlight (Johnson 82). The dual nature of coal reverberates as social bifurcations between oppressed producers and empowered consumers, the respective habitats of a crushing underworld and a rapid-transit surface world, and the collective conscious of a people split between public pride of modernity and a private, often suppressed horror at the tolls exacted on the modern mind and body.

IV Do the locomotion

The coal-fired steam engines of stationary industry led to the reorganization of the workplace in the early nineteenth century, but steam engines drove a more sublime revolution in transportation industry soon after. If stationary industry changed cities, moving industry changed everything in between. The channels along which people traveled by land shifted from muddy, pocked turnpikes to smooth rails, and by sea from meandering trade winds to shortest linear crossings. The advantages of this revolution of moving industry included time saved, more accurate schedules, larger and heavier hauling capacity, and a broader access to transportation across social classes. The disadvantages directly correlated to the advantages: the psychological vertigo from the rapid pace of

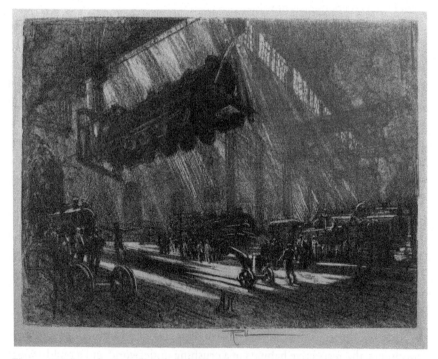

Figure 5.4 Joseph Pennell, *The Flying Locomotive*, 1917. Image courtesy of the National Gallery of Art. Public domain.

commerce, the imposition of a merciless time grid blind to diurnal and seasonal cycles, greater demand for consumable goods and the raw materials that make them, and the inundation of pristine natural places as the new middle class earned vacations and exercised their upward mobility by traveling.

Before the railway, the fastest anyone could travel by land was at the galloping speed of a horse—about twenty-five miles per hour—and this speed was unsustainable for much more than a mile. A more realistic pace in the horse-and-carriage days was about five miles per hour, with a total daily distance of fifty to seventy miles. A speedy mail coach might travel over one hundred miles in a day over long hours with several changes in horses while carrying a few thousand pounds of riders and cargo. The fuel was hay and oats—classic agrarian biomass. The locomotive and its coal fuel initiated an absolute paradigm shift. Within a generation, anyone from commoner to queen could experience speeds of over fifty miles per hour, and could traverse all of England within a day. Not only were locomotives moving steam engines fired by coal, but also steam engines that transported coal, traveling from the British Midlands to industrial textile centers like Birmingham and Leeds and population centers like London.

The first true steam-engine railway journey took place in 1804 at an ironworks manufactory in Wales, which soon became the center of high-pressure steam-engine innovation. The leading spur to the innovation of railways was not originally passenger transport, but hauling coal from mining districts to waterways and manufacturing centers. Pit owners often paid egregious sums to wealthy landowners whose tracts lay between pitheads and nearby rivers. The railway's advance required land purchases that led to legal imbroglios when frequent objections came from those lords benefiting from wayleave. By 1825, locomotives hauled coal, and passengers as well, on the twenty-six-mile Stockton and Darlington Railway in northeast England on a route that carefully avoided the Earl of Darlington's fox holes. The industrial era's iron skull was butting up against the jeweled crown of feudal entitlements.

The modest Stockton and Darlington Railway was soon succeeded by the more important Liverpool and Manchester Railway, which also started as a coal-hauling enterprise but soon turned to passengers. By mid-century, over seven thousand miles of rail traversed England and Scotland. In 1830, the Tom Thumb steam engine was running rails in Maryland, the first stretch of the great American railway that would span from the Atlantic to the Pacific, Canada to Mexico, by century's end.

The first woman to ride the passenger railway left an enduring account of the sublime new feeling of rapid transport on the Stockton and Darlington line.

Referring to the locomotive as a "curious little fire-horse," the actress Fanny Kemble wrote in 1830:

> She goes upon two wheels, which are her feet, and are moved by bright steel legs called pistons; these are propelled by steam, and in proportion as more steam is applied to the upper extremities (the hip-joints, I suppose) of these pistons, the faster they move the wheels; and when it is desirable to diminish the speed, the steam, which unless suffered to escape would burst the boiler, evaporates through a safety-valve into the air. The reins, bit, and bridle of this wonderful beast—a small steel handle, which applies or withdraws the steam for its legs or pistons, so that a child might manage it. The coals, which are its oats, were under the bench, and there was a small glass tube affixed to the boiler, with water in it, which indicates by its fullness or emptiness when the creature wants water, which is immediately conveyed to it from its reservoirs. There is a chimney to the stove, but as they burn coke there is none of the dreadful black smoke which accompanies the progress of a steam-vessel. This snorting little animal, which I felt rather inclined to pat, was then harnessed to our carriage, and Mr. Stephenson having taken me on the bench of the engine with him, we started at about ten miles per hour. (Qtd. in Adams 10)

The engineer George Stephenson, a gruff man by reputation, is in Kemble's eyes "the master of all these marvels," and she is "most horribly in love" with his clipped northern accent and the dour aspect of the romantic hero (qtd. in Adams 13). His passenger appreciates the sensations of rail travel, the "flying white breath and rhythmical, unvarying pace" of the strokes down the line (qtd. in Adams 11). Kemble notes the engineering surrounding the iron horse, including the path along "rocky walls, which are already clothed with moss and ferns and grasses ... great masses of stone had been cut asunder to allow our passage" (qtd. in Adams 11). She implies that the incision of rails into the skin of nature is soon healed with a pleasing secondary naturescape of greenery that softens the cut rocks, a symbiosis of built and natural landscapes. The railway is not smoky, and it is green. The fifteen-mile journey crosses a swamp and a deep valley that had presented monstrous engineering challenges. The line was laid by filling the swamp with tons of hand-hewn rocks supporting a rail platform, and a grand stone viaduct leapt across the Skerne river valley (Wolmar 15). Eventually Stephenson opens up the engine to

> its utmost speed, thirty-five miles an hour, swifter than a bird flies (for they tried the experiment with a snipe). You cannot conceive what that sensation of cutting the air was; the motion is as smooth as possible, too. I could either

have read or written; and as it was, I stood up, and with my bonnet off "drank the air before me." The wind, which was strong, or perhaps the force of our own thrusting against it, absolutely weighed my eyelids down. When I closed my eyes this sensation of flying was quite delightful, and strange beyond description; yet, strange as it was, I had a perfect sense of security, and not the slightest fear … [as] this brave little she-dragon of ours flew on. (Qtd. in Adams 12–13)

Carried above mundane transportation, Kemble's account revels in supernatural wonder at the miraculous technology. Even her bonnet is drunk on the sensation of moving. It is the alchemy of modernity, and Stephenson holds the philosopher's stone. Out of the heavy, dull elements of iron and coal come the floating, sparkling impressions of a woman in flight, defying gravity, rocky obstructions, and time itself as she hurtles through space and into the future with "a perfect sense of security" (qtd. in Adams 13). She is blinded literally by the wind velocity, but also blinded in enchantment by the force and speed of the "she-dragon" and the machine's stern maker. The slow biomass plod is instantly forgotten in the new rail-running world.

The September 1825 opening of the Stockton and Darlington received worldwide attention, and though some Luddite journalists questioned why anyone would want to ride in "a coal wagon, along a dreary wagon-way, and to be dragged, for the greater part of the distance, by a roaring steam-engine," other newspaper editors accurately envisioned the economic significance of the invention (qtd. in Wolmer 12). The editors of the *Newcastle Courant* saw that the railway would "open the London market to the collieries of Durham as well as facilitate the obtaining of fuel to the country along its line" (qtd. in Wolmer 24). With access to fuel came industry, civic development, and exponential increases in the population of the rural hamlets neighboring the rails.

First impressions like Kemble's formed the pro-rail rhetoric of the early nineteenth century. Young Queen Victoria arranged for her own Royal Train to transport her from Windsor Castle to London in 1842, and in 1901 the then-ubiquitous train bore her body out of London for burial. Still, the magnificent iron horse did not ride onto the scene unopposed. Debates in Parliament in the first half of the nineteenth century included irrational fears that the cows would cease to give milk, hens would go off-lay, and useless horses would go extinct (Henderson and Sharpe 1050). The debates also voiced the rational fear that the process of laying down a rail network would irrevocably alter the landscape and the deep-seeded way of life as hundreds of new stations shuffled people by the millions between city and country. With the railway, commerce was no longer confined to the industrial centers, but could cluster around every station on the line—a dream for developers and a nightmare for the conservationists of the day.

Coal-driven trains enabled the first long-distance commutes between cities and the country, but they also opened the portal to far-flung tourism and could carry the masses more effectively than the traveling chaise and four. In the pre-locomotive England of *Pride and Prejudice*, for example, Elizabeth Bennet's pleasure tour to the Lake District with the Gardiners is curtailed by time: due to Mr. Gardiner's demanding business schedule, the original month-long trip is shortened to three weeks. Their itinerary becomes serendipity, as they only go as far as Pemberley in Derbyshire. The difference in distance seems trivial to our fossil-driven sensibilities: it is about 130 miles from the Bennet home in Hertfordshire to Derbyshire, and about three hundred miles from Hertfordshire to the Lake District. The greater distance could be traversed easily in a day using coal or petroleum. But Elizabeth, traveling on the power of oats and roadside grasses, could not manage such a circuit in three weeks on an average of thirty miles per day. She owes her chance reunion with Darcy to the limitations of biomass, and the sexy comedic ending of *Pride and Prejudice* might have been an old-maid tragedy had the novel been set a generation later. (Then again, a Victorian-era Elizabeth might have met a serious young industrialist of no particular family, as Margaret Hale beguiles Mr. Thornton in Gaskell's *North and South*.)

The Gardiners' original vacation plans follow a well-worn path in Austen's time. In the early days of industry, William Wordsworth lamented that his picturesque home in the Lake District was overrun by tourists. In the 1800 poem "The Brothers" he writes,

> THESE Tourists, heaven preserve us! needs must live
> A profitable life: some glance along,
> Rapid and gay, as if the earth were air,
> And they were butterflies to wheel about
> Long as the summer lasted: some, as wise,
> Perched on the forehead of a jutting crag,
> Pencil in hand and book upon the knee,
> Will look and scribble, scribble on and look.

> (ll. 1–8)

To mock the diversions of the seasonal peripatetic was easy enough; the modest and hard-won lifestyles of native farmers, shepherds, and tradesmen stood starkly against the silly pleasure seeking of this "Son of idleness" floating atop the "profitable life" of city commerce. Samuel Taylor Coleridge lampooned the cycles of the "Gold-headed cane on a pikteresk toor," though both poets were themselves strongly influenced by William Gilpin's philosophy of the picturesque, and had conducted their own tour in the autumn of 1799. To the serious young poets, these new tourists had embarrassing accents, conspicuous baubles, and airy heads to match the cloud-blown scenery reflected in the lakes. Jane Austen was also amusingly annoyed by this vogue of picturesque tourism, and protected Elizabeth Bennet from such silliness by alluding to her cultural awareness: "Lakes, mountains, and rivers shall not be jumbled together in our imaginations; nor when we attempt to describe any particular scene, will we begin quarreling about its relative situation. Let *our* first effusions be less insupportable than those of the generality of travellers" (*Pride and Prejudice* 160).

Wordsworth held out his no-tour stance for several more years before he agreed, in 1810 when he was strapped for cash, to write an introduction and narrative for Joseph Wilkinson's *Select Views in Cumberland, Westmoreland, and Lancashire*. By 1820, Wordsworth had published his own *Guide to the Lakes* to profit from the new-money urbanite's hunger for natural scenery at the beginning of industrial-era ecotourism. They might refer to his *Guide* for Lake District superlatives (it is superior in beauty to the Alps and Scotland); for suggestions of picturesque walking and carriage paths; and even for self-chastening commentary on the changes in landscape imposed by the flow of tourists. Even before locomotives opened up the channels of mass transportation in England, stationary industry generated disposable income that buoyed a new industry of pleasure travel to the former backwaters of the kingdom.

Count William Wordsworth among the ambivalent. He did not like commerce, the "getting and spending" that he called a "sordid boon," and which he coped with primarily though escapism. His lifelong infatuation with preindustrial nature, worried mainly by glowering cliff faces and ransacked hazelnut bowers, would be tested acutely in his later adulthood as the railway made inroads to the Lake District. Before he became an elderly poet laureate in 1843, the middle-aged Wordsworth attempted to cultivate a positive perspective on the incursions of coal-driven transportation. In "Steamboats, Viaducts, and Railways," written in 1833, his begrudging acceptance is evident: "Motions and Means, on land and sea at war / With old poetic feeling, not for this, / Shall ye, by Poets even, be judged amiss!" (ll. 1–3). We might paraphrase that clipped tercet: "I withhold judgment on my belligerent adversary." He continues:

> Nor shall your presence, howsoe'er it mar
> The loveliness of Nature, prove a bar
> To the Mind's gaining that prophetic sense

Of future change, that point of vision, whence
May be discovered what in soul ye are.

(ll. 4–8)

These first eight lines, an Italian sonnet's octave, establish the war of the abstract
"Motions and Means" against poets, their feelings, and their subject, nature. The
negatives, nots and nors, are aporias designed to hold together his senses so that
rage does not obliterate the higher mind's capacity to gain a vision and a sense
of the beastly machine. He grants the possibility that the industrial future has
a soul, though incomprehensible to the old-souled nature poet. Now blind, he
hopes eventually to see. The sestet, a hopeful volta, turns to new terrain in the
poem itself and its subject:

In spite of all that beauty may disown
In your harsh features, Nature doth embrace
Her lawful offspring in Man's art; and Time,
Pleased with your triumphs o'er his brother Space,
Accepts from your bold hands the proffered crown
Of hope, and smiles on you with cheer sublime.

(ll. 9–14)

Space, and what it contains (the beauty of nature), have been scarred and
defeated. Still, as he often said in *The Prelude*, man is a child of nature, so the arts
of man can be naturalized by the self-excusing logic that has enabled so much of
industrial development to proceed unfettered. Everything a child of nature does
must be "natural." And time, that hooded lunatic with a terrifying scythe, thinks
it is wonderful that the next generation can obliterate space. People rushing here
and there across three dimensions are less attuned to the passage of the fourth,
though they may be more strictly aware of the current position on the clock
face. Time is crowned with hope, a dangerous positive emotion, and his smiles
are sublime, the most awesome of aesthetic conditions. Sublime is not cheery, it
is fear-y.

By 1844, there had been plenty of time for the mind to gain "that prophetic
sense" of what was to come. Railway mania had arrived and the bubble was high.
Coal fields in the Midlands poured their fuel onto waiting cars. The new queen
chose the railway over her father's equipage. The railway was coming to the Lake
District via Kendal and Windermere. This time it was personal. Wordsworth was
an anointed laureate and his voice spread across the kingdom via the pages of the
Morning Post: "Is then no nook of English ground secure / From rash assault?"

(ll. 1–2). He bemoans the "blight" that will kill his "Schemes of retirement sown / In youth, and 'mid the busy world kept pure" as so many flowers spoiled on "his paternal fields at random thrown" (ll. 2–3, 8). In this protectionist stance, satisfied with its own sense of paternal dominion against the claims of the wider, and presumably sullied, interests of the nation, he bemoans the "false utilitarian lure" of industrial progress that threatens custom with "ruthless change" (ll. 7, 6). The sestet following this bile-filled octave invokes wild nature to rebel against the arts of man:

> Baffle the threat, bright Scene, from Orresthead
> Given to the pausing traveller's rapturous glance:
> Plead for thy peace, thou beautiful romance
> Of nature; and, if human hearts be dead,
> Speak, passing winds; ye torrents, with your strong
> And constant voice, protest against the wrong.
>
> (ll. 9–14)

The station is set to be built at the bottom of Orrest Head, a hill that offers a fine prospect of Lake Windermere. Admirers of the picturesque, part of an industry that Wordsworth helped to create with his *Guide to the Lakes*, merely pause in rapture at the scene, and then move on. A walking tour of it is described minutely in the 1846 edition of the guide. The strict timetables of rail travel allow only a superficial glance, and the next batch of glancers will soon arrive on the next train. Pleading for prosopopoeia, Wordsworth asks the scene itself to baffle, plead, speak, and protest the laying of the rail. Ironically, the more the valley displays its value in beauty, the more enticing the new line that would bring city-weary workers out into the country. Wordsworth wants industrial blight to remain confined to the cities, along with their workers. But Wordsworth's *cri de coeur* was ignored. Royal assent to construct the railway came in 1845, and by 1847 the line ran swiftly past the banks of Lake Windermere carrying hordes of travelers past the natural treasures they had never seen before.

Pleasure travel by train modified itineraries according to rail routes, greatly extended ranges, and welcomed poorer and working people. It also changed conceptions of space, time, and perception. In his inquiry on "The Moral of Landscape," John Ruskin suggests that rapid rail travel is "a mere passing fever" for the childish mind (*Works* Vol. 3 439). To Ruskin, the self-aggrandizing conquest of space is just another indication of a society that is losing touch with value in preference for sensation. True perception requires depth, not breadth: "No changing of place at a hundred miles an hour will make us one whit stronger, or happier, or wiser. There was always more in the world than man could see, walked they ever so slowly; they will see it no better for going fast. The really precious things are thought and sight, not pace." The industrial optimist might see broader horizons thanks to rail-borne travel, which brings access to more of the world. Ruskin sees a world miserably contracted by rapid transit: a world not just smaller but also shallower. The power-seeking fool perceives space and time as inconveniences he wishes to "shorten" and even "kill"; the wise man wants "first to gain them, then to animate them" (320).

Venice, Ruskin's jewel of architecture, is endangered by this time-killing mentality. Noting the Victorians' "impotent feelings of romance" that "gild" the forms of earlier ages, modern tourists liberated by rail travel see only the surface of things (*Stones of Venice* 4). They miss both the depth of the city's degradation and the ecstasy of its enduring beauty:

> Although the last few eventful years, fraught with change to the face of the whole earth, have been more fatal in their influence on Venice than the five hundred that preceded them; though the noble landscape of approach to her can now be seen no more, or seen only by a glance, as the engine slackens its rushing on the iron line; and though many of her palaces are for ever defaced, and many in desecrated ruins, there is still so much of magic in her aspect, that the hurried traveller, who must leave her before the wonder of that first aspect has been worn away, may still be led to forget the humility of her origin, and to shut his eyes to the depth of her desolation. They, at least, are little to be envied, in whose hearts the great charities of the imagination lie dead, and for whom the fancy has no power to repress the importunity of painful impressions, or to raise what is ignoble, and disguise what is discordant, in a scene so rich in its remembrances, so surpassing in its beauty. But for this work of the imagination there must be no permission during the task which is before us. (4)

Ruskin was tremendously wrong when he suggested that train travel and the telegraph were temporary fancies of an over-energized fossil-fuel society still in its childish stage. Arguably, he was also wrong in claiming that rapid travel stifles "seeing" a landscape—it opens more landscapes to be seen, but only to the properly attuned traveler who exits the vehicle. In this late stage of industrialism, fossil fuel is the central commodity of society, fueling a global airline industry that reduces the earth to a forty-five-hour circumference, unites regions with highways and rails, and charges the smart phones obsolescing in the pursuit of profit. Ruskin's philosophy of slow time and deep place is still cultivated in the bioregionalism and slow food movements; these countercultures exhibit an alternative ontology to the industrial pace race.

Coal-chuffing locomotives were perhaps most notable for their proficiency in opening rural and wild districts to the influx of commerce. Manufacturing and tourist industries had bewildering effects on the formerly quiet provinces, which were cut through with rails joining productive cities. The first train horn ringing the hills was an aural symbol; once it was heard, change was irrevocable. Because of the stunning single-generation shift in the countryside from a deep medieval past to a modern industrial future, other landscapes altered by moving industry get less attention. Urban spaces were also changed by coal-fired transportation. Citing the destructive construction of the London and Birmingham railway, which opened in 1838, Charles Dickens describes the ruin brought upon London's Camden Town that he knew as a boy. In *Dombey and Son*, he writes of a "great earthquake" that "rent the whole neighborhood to its centre" as though railway construction were a kind of primordial chaos:

> Houses were knocked down; streets broken through and stopped; deep pits and trenches dug in the ground, enormous heaps of earth and clay thrown up; buildings that were undermined and shaking, propped by great beams of wood. Here, a chaos of carts, overthrown and jumbled together, lay topsy-turvy at the bottom of a steep unnatural hill; there, confused treasures of iron soaked and rusted in something that had accidentally become a pond. Everywhere were bridges that led to nowhere; thoroughfares that were wholly impassable; Babel towers of chimneys, wanting half their height ... piles of scaffolding, and wildernesses of bricks, and giant forms of cranes, and tripods straddling above nothing.... Hot springs and fiery eruptions, the usual attendants upon earthquakes, lent their contributions of confusion upon the scene. Boiling water hissed and heaved within dilapidated walls; whence, also, the glare and roar of flames came issuing forth; and mounds of ashes blocked up rights of way, and wholly changed the law and custom of the neighborhood.
>
> In short, the yet unfinished and unopened Railroad was in progress; and, from the very core of all this dire disorder, trailed smoothly away, upon its mighty course of civilisation and improvement. (68)

Dickens's absurdly over-constructed image of deconstruction derides the diminishing returns of industrial "improvement." With the verbal equivalent of an earthquake, Dickens shows the violence of retrofitting an ancient city with a modern railway. His invocation of chaos and hell, volcanoes and earthquakes, and Babelian miscommunication counters the progressivism of railway boosters. Their project annihilates past and present, "law and custom," in its singular purpose to lay a straight track to the future. Quite the opposite of Fanny Kemble's ode to the railroad, which celebrates a rapid, smooth, fragrant, green, mossy path into the future, Dickens's contemporary voice cries havoc and wastage. Sublime heaps of history lie in rubble along the future's survey route.

Dombey and Son was published in 1846. By 1854, Dickens saw fit to write a book that concentrated entirely on the cruelty of industrial thinking and disgust for the living conditions in industrialized England. In *Hard Times*, the hapless factory workers of Coketown live "by the railway, where the danger-lights were waning in the strengthening day; by the railway's crazy neighbourhood, half pulled down and half built up; by scattered red brick villas, where the besmoked evergreens were sprinkled with a dirty powder, like untidy snuff-takers; by coal-dust paths and many varieties of ugliness" (155). Echoing Engels's account of Manchester's environment, Dickens extends the industrial pariah from stationary to moving industry; proletariats suffer in the shadow of both steam engines.

Not so for the barons of Coketown, the "national cinder-heap," who are the first generation of suburbanites fleeing the pollution and crowds of the city via subsidized public transportation: "Mr. Bounderby had taken possession of a house and grounds, about fifteen miles from the town, and accessible within a mile or two, by a railway striding on many arches over a wild country, undermined by deserted coal-shafts, and spotted at night by fires and black shapes of stationary engines at pits' mouths" (192, 157). The land surrounding Coketown is strangely used: it is undeveloped, but already degraded by the disused shafts and ongoing pit mine operations signaled by puffs of coal smoke. Bounderby's suburban estate turns its backside to those dirty associations that make its existence possible. His house looks instead toward the mellower "rustic landscape, golden with heath, and snowy with hawthorn in the spring of the year, and tremulous with leaves and their shadows all the summer time" (157). This pastoral idyll, conveniently accessed by rail to the remunerative city, foreshadows the suburban settlement pattern distinctive to the mid-twentieth century, where every day a city worker could travel back in time to the illusion of a pastoral existence. Mr. Bounderby's unctuous companion Mrs. Sparsit, catching the train from town to country, "was borne along the arches spanning the land of coal-pits past and present, as if she had been caught up in a cloud and whirled away" (195). Like Fanny Kemble, the imagination of Mrs. Sparsit takes flight while traveling the grand spans of the railway, sheltered from the unhappy realities—such as coal—that are conditions of its existence. "It afforded Mr. Bounderby supreme satisfaction to instal himself in this snug little estate, and with demonstrative humility to grow cabbages in the flower-garden" (158). In Dickens's sarcastic "humility," the decidedly modern and urban Bounderby gently tends cabbages like a peasant in order to unwind after hours of rancor in the factory. He has the privilege to live two lives, and a railway that links them together.

In *Hard Times*, Dickens allows the poor workers to use the railway for fleeting afternoons of leisure as well. Since it is an instrument of social equality, affording transportation to anyone with fare, the workers play the idyll for an afternoon: "It was customary for those who now and then thirsted for a draught of pure air, which is not absolutely the most wicked among the vanities of life, to get a few miles away by the railroad, and then begin their walk, or their lounge in the fields" (245). Sissy and Rachael ride out to a station halfway to the master Bounderby's estate, and their escape is possible on a Sunday, since the Coketown soot climate is not yet universal. Nature abides as if nobody had ever heard of mass-produced tweed: "Though the green landscape was blotted here and there with heaps of coal, it was green elsewhere, and there were trees to see, and there were larks singing (though it was Sunday), and there were pleasant scents in the air, and all was over-arched by a bright blue sky" (245). The women gain perspective, literally. They see "black mist" in the direction of Coketown, hills in another direction, and in the third a gilded horizon lighting on the ocean (245).

The scene is quiescent and atavistic; they have stepped backward in time: "Engines at pits' mouths, and lean old horses that had worn the circle of their daily labour into the ground, were alike quiet; wheels had ceased for a short space to turn; and the great wheel of earth seemed to revolve without the shocks and noises of another time" (245). What is pleasant at first becomes creepy as they traverse the abandoned countryside, which appears as a ghost land in the wake of population flight to urban manufactures. Ghost pastoral turns to sublime disaster zone as they discover Stephen Blackpool fallen into a "black ragged chasm hidden by the thick grass" aptly named the Old Hell Shaft (246). The holes in the countryside are not only the vacancies left by the lost economy and society of rural England but actual fissures in a landscape stripped of coal and left unrestored. Blackpool's life ends in his eponymous place, swallowed by the mine, a sacrifice to fuel.

The death that Dickens designed for Stephen Blackpool is highly symbolic. The factory hand is bodily swallowed by the very tomb excavated to fuel his work. The worker is both a figurative and literal food gobbled up by the industrial system. Coal's physical advantages as a fuel are cancelled out by the social and natural conditions it creates: bloated taskmasters, winnowed workers, and the shoddy morality of industrialism's ethos of the self. This lugubrious economic order is built upon the seeming absurdities of filth-filled cities, coal country riddled with death holes, and a modern train flying over the squalor on a whimsical viaduct bearing its master to his idyllic estate. The tragedy is that the absurd is a true reflection of coal's Victorian reality.

More measured novels of the industrial Midlands, like Gaskell's *North and South*, take account of the opportunities opened by hard-working industrialists as well as the ugliness of that new world. Like Dickens, Gaskell overtly compares manufacturing centers with bucolic backwaters, and crafts a woman's happiness out of her disavowal of the old rural life in preference for the industrial nexus. Margaret Hale would not be able to perceive the relative virtues of north (industry) and south (agronomy) without her seat on the railway. She not only finds her husband among the machines; she finds her own proactive place in a world that permits some new opportunities for women: she advocates between workers and mill masters.

The whole novel is in sync with the railway table as characters come and go from far-flung locales, and the cinema version depicts the first kiss between the betrothed couple in public on a train station bench, before they retreat to the privacy of a jostling train car. Coal transportation created the modern sense of the world as small and accessible, permitting an egocentric mastery over time and space coupled with the voyeuristic eye of the passenger gazing through a picture window. It empowered the individual body to move at superhuman pace through space.

Inversely, and simultaneously, coal disempowered millions of workers from the East and West Indies to western Europe as its new world order demanded raw materials produced by slaves and textiles produced by proletariats. It was also an ingenious shifting-scene-maker for writers: the train car trumped the horse-drawn carriage as a new setting for mischief. Patricia Highsmith would unite murderers in her 1950 novel *Strangers on a Train*, and Agatha Christie in the 1957 novel *4:50 from Paddington*. With the train car as an unreliable, shifting location, writers could play tricks with time and space and attenuate the nature of witness. From a solid, Newtonian world of predictable movement there emerged the more harrowing experience of relativity, with fast-running trains putting into question the physical nature of simultaneity and the observer's frame of reference.

The railroad was arguably even more significant to the development of North America than it was in England. With much more and much wilder terrain, less infrastructure, greater natural resources, and lower population density, the building of the great American railway was needed to unify the vast territory. Its construction was generally viewed with less controversy because, west of the East Coast plain, the railway created the possibility of large-scale settlement and trade. In England, everything had to be retrofitted over deep-seeded properties and traditions. In America, the new rails shot through vast unpopulated areas, and preexisting cultures endangered by its advance—namely, Native American tribes—had already been decimated and marginalized by the middle of the nineteenth century. In 1869, the line from San Francisco met the line from New York midway at Council Bluffs, Iowa, completing the first transcontinental railway and forever changing the aura of wilderness in North America. The frontier was bisected.

In American literature, the locomotive is generally an object of reverence. Walt Whitman's *Song of Myself* captures this spirit of celebration of the technology and fuel that enabled the opening of the American continent. Always embodied and frequently sexual, Whitman's poetry echoes the bawdy ballads of industrial London. In "To a Locomotive in Winter," the machine is a "fierce-throated beauty" whose fire and song dispel the cold-closing night:

> Thee in thy panoply, thy measured dual throbbing, and thy beat convulsive;
> Thy black cylindric body, golden brass, and silvery steel;
> Thy ponderous side-bars, parallel and connecting rods, gyrating, shuttling at thy sides;
>
> Thy metrical, now swelling pant and roar—now tapering in the distance;
> Thy great protruding head-light, fix'd in front;
> Thy long, pale, floating vapor-pennants, tinged with delicate purple;
> The dense and murky clouds out-belching from thy smoke-stack;
> Thy knitted frame—thy springs and valves—the tremulous twinkle of thy wheels;
> Thy train of cars behind, obedient, merrily-following,
> Through gale or calm, now swift, now slack, yet steadily careering:

(ll. 3–12)

The machine is sensual and superhuman in size and power. Its dragon-like might shrinks vast distances, cuts through the cold, and brings people and goods "merrily" along on this great modern ride. Fanny Kemble's delight at moving thirty-five miles per hour is upgraded to epic speed and distance in any weather; Whitman sings the locomotive as a new archetype of modern agency. Whitman explodes the old literary comforts—"No sweetness debonair of tearful harp or glib piano thine"—to sound his barbaric yawp on behalf of the mighty engine, "Type of the modern! emblem of motion and power! pulse of the continent!" (ll. 21, 13).

The industry-friendly encomia of Kemble and Whitman have been recuperated in contemporary environmental arguments for the new roles of the railway in a low-carbon industrial world. There is irony in hailing that prime mover of yesterday's industry as tomorrow's environmental savior, as the solution that will replace the private car and reunify sprawled communities along a corridor. After all, fed on coal, the locomotive was first used to haul coal from mines. Within a few decades it also disgorged the masses into environmentally delicate locales; facilitated mass consumerism by hauling freight between remote factories and urban retailers; and elided the distinction between access and exploitation, as it opened the American and British colonies to the demands of the metropole for natural resources like timber, minerals, and game animals. The rail corridor became a vector for invasive species that were sometimes dropped as seeds by the leaky metal cars, but more often crept out of endemic locales to grow along the disturbed habitat of the trackside (Revill 222–3).

How could the locomotive be framed as an environmental advantage? The main answer, both sensible and short, is that the railway is stunningly more efficient in its use of fuel per weight transported. On average, a train can move one ton of freight 473 miles using one gallon of fuel, and the Association of American Railroads claims that trains are about four times more efficient than the standard trailer trucks that fill the highways ("Environmental Benefits" 1). Their wider adoption would ease road congestion from truck traffic, presumably allowing greater access for private cars (but that is not the point, right?). Modern railways present themselves as people-friendly, too: digital-age white-collar workers are attracted by Wi-Fi access, space to work and socialize, and freedom from the stresses of car ownership and the road rage of rush-hour driving. The railway's long, skinny habitat can be cultivated as a multiuse terrain for biking, animal migration, wildflower habitat, and even small-scale gardening. Railways are enjoying a renaissance amid an aging and overcrowded highway system. Their appeal is at once nostalgic and chic-futuristic, practical and fantastical.

In the infatuation of modern train love, it should not be forgotten that their original purpose continues to be a major part of their work. Trains haul coal to the power plants that generate electricity. Trains haul crude and shale oil to refineries, and recent spectacular disasters like the one that killed forty-seven people in Lac-Megantic, Quebec in 2013, underscore the danger of dragging vats of fossil fuel along the iron road. Down to its belly, the boiler, the locomotive is an industrial beast of burden continually beholden to fossil fuels. The railway is more efficient than the highway, but its very efficiency tends to exacerbate carbon emissions by linking the producers of fossil fuels with its worldwide consumers.

Coal walked on water as well. Steamships proved their commercial viability nearly as early as the steam-powered locomotive: in August 1807, the steamship *North River* (aka *Clermont*) first navigated up the Hudson River from New York City to Albany, leaving dozens of hard-tacking sailing ships in its wake. Regular passenger service began the next month, in September 1807, and the ferry service turned an immediate profit. The steam-powered ocean liner is an ideal figure for the divergent fortunes of the industrial social classes: onboard were perpetuated the old distinctions of rank that were deployed on shore, based upon conditions of transport—like foot versus horseback, hackney coach versus barouche, and third versus first class on the railways. Whether biomass or fossil fueled, passengers got what they paid for. The ocean liner mapped these social divisions onto the geography of the ship, with the highest payers on their posh decks, above the rest of the passengers, and the poorest voyagers sardine packed in steerage, between the cables controlling the rudder and below water level.

As in the ceaseless toil of stationary industry, the workers shoveling coal in the steamship were conscripted into a machine-like existence, forever filling the rapacious maw of the great engines that carried thousands of tons across the oceans. Rather than figuring these men as post-human cyborgs, both man and machine, modernist literature often responded to fears of biological degeneration by capturing them in atavistic forms: coal tenders became apes made bestial by their work. Evolutionary theory had created a discourse of atavism and degeneration, as the world came to see human history in continuum with other animals rather than in static separation from them. The unforgiving task of shoveling coal created an anachronistic savage man adapted to the industrial jungle of the boiler room.

Eugene O'Neill called this man the "hairy ape," and unsubtly fixed him in an eponymous play (1922) where the human ape, Yank, begins the action flush with pride as a prime mover in industrial society. After a fleeting encounter with a delicate heiress, whose curiosity about the conditions of the working class lead her to the engine room of a great ocean liner, Yank's rage at her horror of him leads to series of failures to integrate into groups: laborers, socialist workers, and prisoners. He ends up dying in the crushing arms of an ape in the zoo. Part of the resemblance comes from the physical nature of the work in an age when the working classes performed repetitive machine-driven tasks that led to hypertrophied physiques. Yank and his compatriots have, in O'Neill's stage directions, a "natural stooping posture which shoveling coal and the resultant over-development of back and shoulder muscles have given them. The men themselves should resemble those pictures in which the appearance of Neanderthal man is guessed at. All are hairy-chested, with long arms of tremendous power, and low, receding brows above their small, fierce, resentful eyes" (2). Their overdeveloped bodies dominate their underdeveloped brains, and their lack of humanity is located in a cognitive–emotional deficit of kinship: the hatred of the abject man. O'Neill's play explores how the subaltern bodies of fossil-fuel workers were disproportionately taxed in order to provide privileged classes with fossil-fueled liberation (Johnson 57). In this order of things, the hysterically feminine heiress enjoys the mind–body pleasure of modern power, which extends her own modest form beyond fleshly bounds to the prosthetics of the mighty ship and the laboring, lumbering bodies that make it go. Yank is a social inferior, stigmatized by social Darwinism into a merely animal, muscular organ; his body itself is owned and used up by her class of consumers—he might as well be coal.

Yank is the most exaggerated of a team of laborers who maintain some individuality based on nostalgia and collective social activism. The elderly Irishman Paddy recalls the bygone work of sailing ships in a celestial symphony of pure air, sun, stars, and exotic coasts. "A warm sun on the clean decks. Sun warming the blood of you, and wind over the miles of shiny green ocean like strong drink to your lungs. Work—aye, hard work—but who'd mind that at all? Sure, you worked under the sky and 'twas work wid skill and daring to it. And wid the day done, in the dog watch, smoking me pipe at ease, the lookout would be raising land maybe, and we'd see the mountains of South Americy wid the red fire of the setting sun painting their white tops and the clouds floating by them!" (18). Like Ishmael, Paddy is eloquent on the idealisms of preindustrial labor lost, but none of these sailing pleasures are afforded the coal scuttlers on the New York to Southampton line. Paddy perceives "black smoke from the funnels smudging the sea, smudging the decks—the bloody engines pounding and throbbing and shaking—wid divil a sight of sun or a breath of clean air—choking our lungs wid coal dust—breaking our backs and hearts in the hell of the stokehole—feeding the bloody furnace—feeding our lives along wid the coal, I'm thinking—caged in by steel from a sight of the sky like bloody apes in the Zoo!" (18). The socialist, Long, blames their desperate labor conditions on an oppressive capitalist system: "We wasn't born this rotten way. All men is born free and ekal.... them lazy, bloated swine what travels first cabin? Them's the ones. They dragged us down 'til we're on'y wage slaves in the bowels of a bloody ship, sweatin', burnin' up, eatin' coal dust! Hit's them's ter blame—the damned capitalist clarss!" (19).

Unwilling to perceive himself as a victim of uncontrollable circumstance, Yank ignores old men and socialists, pigeonholes his pride in the engine room, a synecdoche of the great forces of coal-powered industry of which he considers himself master.

> Sure I'm part of de engines! Why de hell not! Dey move, don't dey? Dey're speed, ain't dey? Dey smash trou, don't dey? Twenty-five knots a hour! Dat's goin' some! Dat's new stuff! Dat belongs! But him, he's too old. He gets dizzy.... Everyting else dat makes de woild move, somep'n makes it move. It can't move witout somep'n else, see? Den yuh get down to me. I'm at de bottom, get me! Dere ain't nothin' foither. I'm de end! I'm de start! I start somep'n and de woild moves! It—dat's me!—de new dat's moiderin' de old! I'm de ting in coal dat makes it boin; I'm steam and oil for de engines; I'm de ting in noise dat makes yuh hear it; I'm smoke and express trains and steamers and factory whistles; I'm de ting in gold dat makes it money! And I'm what makes iron into steel! Steel, dat stands for de whole ting! And I'm steel—steel—steel! I'm de muscles in steel, de punch behind it! (21)

His displaces his anger in aggression toward the coal heaps. Yank's tragic pride plays out in his hubristic narrative of laborer as world maker, the Atlas supporting a ponderous and oblivious globe. Though Yank's speech is at once noble and farcical, it is the truth of his self-vision that testifies to the sheer magnitude of the forces unleashed by coal workers. A coal miner who consumes 3,500 food calories to produce five hundred pounds of coal in a day achieves a stunning gain in energy: he produces five hundred heat calories from coal for every calorie of food burned (Wrigley 206). Yank is a fireman, not a miner, so his nexus falls between the cold lumps and the searing boiler flames, with thousands of calories scuttling from every shovel heave. Though Yank's physique and exertion resemble his ancestors with their shoulder to the plow, his work unleashes the energy of a superhuman by calorie comparison. No wonder Yank feels personal pride despite his imprisonment in the system. He is at once inmate and wizard.

His narcissism is a mirror cracked when the gaze of the steel heiress, Mildred, settles on him in his labor pit. Her stupefied incomprehension of Yank's world reveals to him the great chasm between the benefactors of coal and its slaves. Mildred is coal's paradoxical daughter: translucent white skin, gauzy white dress, weak, protected, and tepid. Yank is coal's prodigal son: massively muscled, arrogant, ignorant, and destined to dig his own grave. The coal world protects its oblivious mistress, and shows no mercy toward its plebeian. Still, the encounter between them in the ship's diabolical engine room is the moment of traumatic awakening for Mildred. Her worldview of universal, easy pleasure ruptures with the knowledge that not only Yank exists, and toils in a coal hell, but that her own life stands on the brutal oppression of the underclasses (Johnson 99). The horror of her epiphany sprawls beyond this memorable scene into O'Neill's maturing twentieth century, where collective middle- and upper-class guilt and anxiety undergirds, and ironically energizes, consumer appetites.

Literature is filled with tales of rich sheltered ladies and mucky laboring men, of their romances and mischances, but O'Neill's play is a modern American romance of hatred, ignorance, fear, and powerless fury. He eschews realism to attain the industrial sublime, his ship seared by white furnace light, shadowed by hulking forms and their unnatural soot dress. Between the antipodes of Mildred and Yank there is a wall built of coal, preserving unequal class distinctions as effectively as the outworn European class system based on land and titles.

V Horizon gone: Mountaintop removal mining

Mountaintop removal mining (MTR) blasts the tops of mountains and fills adjacent valleys with the spoil. Traditional mines opened in the bottoms and amassed spoil at the top. With mountaintop removal, gravity is on the side of the massive earth-moving machines that shove the land above the coal seams, the "overburden," down into the river valleys. Bulldozers skim off the exposed coal, other large trucks and trains haul the product to its markets, and the need to employ miners is greatly reduced from traditional shaft-mine techniques. This technique has refigured the Appalachian landscapes of Ohio and West Virginia most drastically. Advocates of mountaintop removal sketch out a pleasant pastoral future for the sacrifice of the steep forested hollows. By planting grass seed in the valley fill and grading the slope to be flatter, mining engineers envision high meadows that would support game species like elk, deer, and turkey, and with them new hunting areas. They also point to several golf courses built upon reclamation sites. A grassy version of Lake Powell, the water body created by flooding Glen Canyon, this post-mining meadow is a terra-formed landscape. The authors are downright giddy about the environmental benefits of mountaintop removal: "It is clearly evident from the development taking place in the Appalachian region, which is creating jobs, living spaces and recreation for future generations, that this is one of the best examples of sustainable development, not only in the United States, but in the world" (Gardner and Sainato 55).

Ecologists reject this happy depiction of the after-scapes of MTR operations. The chemical impacts on land and water are so fundamental that every living thing that exists in these diverse hollows is affected, and often killed or sickened. The effects are measureable from the smallest phyto- and zooplankton, the base of the food chain, right up through communities of plants and animals, to the humans who have often dwelled for generations in the deep slopes of Appalachia. On a higher level of landscape aesthetics, the evolved fabric of the ridge-and-hollow is torn apart and hastily replaced with a flat pile of rock shambles and overburdened waste lagoons. Or, depending on your perspective, open land for new hunting parks and golf courses.

MTR is now nearly half a century old, with the first operations beginning in the early 1970s. Ann Pancake's novel *Strange As This Weather Has Been* (2007) tells the story of a community in southern West Virginia whose three generations have lived with MTR since its inception. Pancake based her fictional book on interviews and events that occurred in coal communities over the past several decades. She alludes to the Buffalo Creek disaster, still memorialized by long-suffering West Virginians whose lives are ruled by coal: it is, on one hand, their history, tradition, and livelihood and, on the other, coal is the cause of intensifying medical, social, and ecological degradation.

Figure 5.5 Flashdark, *Valley fill—Mountaintop Removal Coal Mining in Martin County, Kentucky*, 2006. Source: Wikimedia Commons. Public domain.

The balance between jobs and quality of life in these Appalachian hollows is tipping, however, toward the worse. Traditional shaft mining employed thousands of workers, providing them with steady jobs and even union representation through most of the twentieth century. Pancake's novel takes place in the late 1990s, when MTR is replacing shaft mining, which kept the surface environment and communities intact. The title is an allusion to climate change in which Pancake links coal burning with "global weirding"—the scrambling of the seasons evidenced in changes to temperature, rain, and cloud cover:

> This year it seemed the seasons were running backwards. The summer strangely cool and wet following a warm snowless winter, that winter following the worst drought summer in sixty years.... too much water or too little, the temperature too high or too low. "Strange as this weather has been," people would say, or, "With this crazy weather we've been having." And I knew Lace believed the weather was linked to the rest of this mess, but I wasn't sure how. (101)

The local weather implies global climate shifts, so Pancake's perspective remains locked within the community. The local Yellowroot Mountain has been bought by the mining company, and the citizens living in its shadow have no power to change the permitting, the MTR technique itself, or even be employed by the company doing the work. They are merely horrified bystanders. The novel depends upon the visual outrage and visceral disgust that surrounds MTR sites, from the bottoms straight up to the truncated peaks. First, from the communities in the bottoms, the floods strip the surface of reformed mountain:

> For a couple hundred yards, it was just the same hollow it had always been, except for the flood mess on the creek.... The creek water itself still colored like creamed coffee left for weeks on a counter.... The sides of the hollow, as we got further in, more naked and scalped, more trees coming down, and up above, mostly just scraggly weeds, the ground deep-ribbed with erosion, and I told myself, yes, this is where the floods come from. From the busted ponds and the confused new shape of the land. From how the land has forgot where the water should go, so the water is just running off every which way. (15–16)

The old slope has been transformed into a mess of organic death and man-made trash caught in a welter between insistent gravity and disrupted watershed dynamics. The floods of muck and poison that inundate the community become as commonplace as the rains themselves, and the protests of the people are met with calculated corporate obfuscation. So the people—like the Luddites of old—break laws and locks, hop gates, and duck security to have a look at the source of the degradation. They find an ecocidal scene:

> The top of Yellowroot was just plain gone. Where ridgetop used to be, nothing by sky.... Then it dawned on me exactly what I was standing under—Yellowroot Mountain, dead. I knew from Lace and Uncle Mogey that after they blasted the top off the mountain to get the coal, they had no place to put the mountain's body except dump it in the head of the hollow. So there it loomed. Pure mountain guts. Hundreds of feet high, hundreds of feet wide. Yellowroot Mountain blasted into bits, turned inside out, then dumped into Yellowroot Creek. (19–20)

The daredevil son, Corey, loves machines and violence. He sees in the mine site an adult-sized playground to show off his ATV tricks. His father Jimmy defends the coal industry, parroting the conventional reasons of jobs and culture, and stays too close to his TV to ever glimpse the new lay of the land. But once Lace, Bant, and Avery break the law to get a glimpse of the violated mountain, they cannot un-see the scene.

Pancake figures the destroyed mountain as a living creature beheaded, drawn, and quartered. Rather than a single death, though, this guillotine is just the beginning of a reign of terror in the community. Those loyal to the coal industry fight with the insurgents against it, causing a schism in the small town. The historical background of meager education and underemployment is exacerbated by losses in property and health problems, as well as the general malaise of living in a place with continual earthquakes from the blasting, and noisy traffic from the heavy machines. The redeeming value of MTR, its coal, condemns the community to oblivion. It generates a little business for gas stations, restaurants, and hotels due the "scabs" that come from out of state to work the operation—but little else. What about the pastoral landscape promised in its wake? Pancake's description of MTR is nothing at all like the brochures of its boosters; in fact it is dystopian:

> He climbs past the first series of terraced ponds, the water as opaque as mustard and colored like the inside of a sick baby's diaper. The only growing thing left up here in the head of the hollow is the grass covering the pond banks, no doubt the same stuff they've genetically engineered for reclamation. Grass that can grow on asphalt. Besides the grass, everything is dead, the hollow an amphitheatre of kill, and the grass itself isn't even green. (213)

When the palimpsest of biodiversity is blown away and the sky looks down on rubble, characters representing three generations—old Mrs. Taylor who predicts

God's apocalypse, Lace the enduring mother/gatherer, and Bant the inquisitive teenager—testify to the human ecological toll of our need for fuel. Mountains are ground down to burn and turn turbines, and what is sacrificed is the fabric, mesh, and web of life that evolved for the five hundred million years since the Appalachians rose out of the crunched-up North American craton. Pancake's novel remains trained on a single community in one hollow, but the implications of this real-life fiction spans the entire region of eastern Kentucky and Ohio and southern West Virginia, where MTR is presently most active. The result, piled on top of all the other things discarded by the industry, is utter disorientation:

> The hardest thing of all about living through this, hasn't been the blasting or the dust or the flooding or the fires or how they broke the community. It's looking up there each morning, at a landscape you had around you every day of your life.
> And seeing your horizon gone. (309)

Pancake's novel brings attention to abject people and natural landscapes that have long been neglected by the modern narrative of progress that features the emancipation of the consumer. The Polyanna narrative neglects accounting for the class and cultural isolation of coal-producing regions, and even sympathetic liberal environmental writing more often relies on privileged subjects, like urban gardening and farmer's markets, read by the bourgeois, rather than attending to the social and environmental traumas suffered in coal-producing regions. Such cultivated myopia about the costs of this fuel are psychologically explicable due to its continuing contributions to the modern life, from domestic heat and electricity to the global manufacture of consumer goods using coal.

The inconvenient truths about coal, like climate change, are traumas suppressed in the cultural conscience that permit the continuing division between coal producers and coal consumers (Johnson 94). This plays out in political spheres, with antienvironmental politicians appealing to coal region voters who feel affinity, economic and cultural, to the working-class narrative of sacrifice and perseverance, even at the cost of their health and habitat. Whether that political story results in steady work or more layoffs due to MTR's low demand for labor is lost in a class-charged imbroglio. The Kentucky writer Wendell Berry has concisely summed up his position on MTR in his region with the holler, "Compromise? Hell! ... Most of us are still too sane to piss in our own cistern, but we allow others to do so, and we reward them for it.... those who piss in our cistern are wealthier than the rest of us" (*Compromise* 155). Ironically, MTR's practice may be endangered not by the entrenched regional opposition or environmental regulation, but by fuel economics. In the United

States, hydro-fracking is making natural gas a cheaper and cleaner alternative to coal, and coal's market is dwindling.

Though the richest veins of literary coal run through the nineteenth century, accounts of air pollution continue to be relevant in the twenty-first century. The miasma has drifted from retrospective industrial England to the developing world's intensive industrial locales, most notably modern China's numerous manufacturing cities, which are choking on coal pollution. London, Manchester, Sheffield, and Leeds have become Beijing, Lanzhou, Guangzhou, Xingtai, Shijiazhuang, Baoding, Handan, Hengshui, and Tangshan. Spirits of the cleansing rain in Swift's "Description of a City Shower," China has the world's most advanced and carefully guarded research on seeding clouds to wash away the smog in its industrial centers. These cities consume coal in order to drive the factories that manufacture consumer goods for the global market, but air pollution cannot be exported, it can only be washed away. China's state-run media has announced that in 2015 the country will begin manufacturing rain to clear away smog using 6,902 cloud-seeding artillery guns and the catalysts of dry ice, silver iodide, and salt powder (Wan).

Gemstone to carbuncle, coal is multitudinous in character and unparalleled in its impact on industrial history. Coal is lead author of the Anthropocene. As literature is a body of ideas and propositions that tends to critique the hegemony, our survey of coal in fiction and poetry more often highlights the sacrifices inherent in our adoption of high-energy economies and lifestyles than it acts as impresario for the miracle of fossil fuels. For the speed, energy, consumables, and infrastructure gained, indeed an entire paradigm shift in human ontology from agricultural craft societies to industrial ones, literature takes a counterpoint account of the mass losses of life and landscape. Still, it preserves some of the wonder inherent in the superhuman experience of movement beyond animal capabilities, a wonder we've mostly lost because of its quotidian nature. What was miracle has become mundane. What was freedom has become fetters—of urban crowds, debt-ridden consumerism, pollution. This cultural evolution from early industry into middle and late forms of industrial capitalism follows a parallel arc along coal's cousin fuel: petroleum.

6

Black Gold

I Oil ontology

Oil pulses through the carburetors of the industrial world. It is the fossil fuel of modern motion. Once oil became widely and cheaply available sometime between the twentieth century's world wars, various grades of diesel and gasoline drove engineering innovation toward the automobile. Old coal drove the locomotives and steamships of nineteenth-century moving industry; now, diesel and heavy fuel oil do that work. As a symbol of freedom, power, and recklessness, the automobile has left its tire tracks on many a literary page. We'll kick those tires and take them for a spin in this chapter.

Fossil oil is not only a fuel of motion; its carbon chains appear in a mind-boggling array of materials. Anything not explicitly metal or glass often has some petroleum component, usually one of the ubiquitous plastics of modern life. Its usefulness is not just convenient, it is equally terrifying. The architecture of late-capitalist consumerism would collapse without load-bearing oil. No computers or television without oil: no Amazon.com, Facebook, OK Cupid, or Candy Crush. No big box stores without oil. No recycling or garbage pickup. No landfills. No clean municipal water. No conventional agriculture, which uses fossil-based pesticides, and provides 95 percent of food sold in America (Greene).

Some claim that "Peak Oil"—the point beyond which supplies begin to diminish—is already upon us. Particularly, the Hubbert peak theory predicts that global supplies will diminish in the early twenty-first century, just as huge populations in India and China come online, intensifying demand. But low-grade sources like Canada's oil sands obscure the supply crisis and introduce an array of disturbing questions about just how desperately we need petroleum. The conversation evolves from "peak oil" to "tough oil." Losing the artificially cheap prices that have floated the twentieth century's big-growth economies seems inevitable. He may not need those stores of water, rice, beans, beef jerky, and ammo your cousin stashed under his cabin after reading Hubbert's theory, but it is certainly worth pondering the glory days, decline, and demise of that pressed phytoplankton you pump into your tank.

Petroleum's creation story is a geological recipe for braised biomass. Crude oil goes through a geological diagenesis broadly similar to coal, in that both involve ancient biomass compressed and buried in oxygen-starved environments. However, crude oil's organic source point is mostly green algae from ancient seas, and coal is the slow-cooked bodies of swamp-side ferns and trees. Plant and animal plankton lived in oceans and, over the course of millennia, numberless hordes of them died, sank, and were covered by the next generation's dead. Deep swales of these sunken plankton bodies were kept in anaerobic conditions. In the next stage of catagenesis, the proto-oil was gently heated for millions of years until it reached the sweet spot between 150 and 300 degrees Fahrenheit (65–150 degrees Celsius). If it got too hot, the ancient plankton were cooked into natural gas.

Because ancient lakes and seas held variable species under an array of conditions, crude oil is now extracted in many forms, from the prized and diminishing "light sweet crude" to the abundant filthy "tar sands" that are, with much controversy, a large share of the petroleum industry's future. Crude oil is useless in raw form; it must be refined by superheating to separate the molecules by size. Crude stocks leave the refinery with only about half the oil in gasoline form, ready for our cars, with small molecules of four to twelve carbon atoms. About 40 percent of the remaining molecules have longer carbon chains that serve our needs as home heating oil, diesel fuel, jet fuel, and kerosene; 5 percent provide petrochemicals for manufacturing plastics and other petro-products like paint, aspirin, and chewing gum; another 5 percent of the heaviest and largest molecules will have a future life as lubricants, greases, and asphalt (Texas Energy Museum).

Literary critics call our modern state an "oil ontology," which is to say that it has become so omnipresent, we can't even see oil because we look through oil glasses, or oil eyes (Hitchcock 81). We are, ourselves, petroleum products, an intermixture of self-generating and manufactured parts. This makes us cyborgs, and not just because some of us have artificial hips and pacemakers made of petroleum parts. Materially, we are part petro because the food we eat is fertilized, harvested, processed, distributed, and sold using petrochemicals. Our car wheels are prosthetics for the legs we less often use; our computers are our hands and voices and the brain-like repositories of our habits and ideas.

The sensuousness of oil assists in its aesthetics, making it alternately alluring and revolting. If earth is a goddess called Gaia, then oil is one of her fluids. When drilled, it coats and permeates its environment in a suffocating black shroud. It is black and smelly like shit, and Venezuelan president Hugo Chavez, rich and politically empowered by the stuff, called it "the devil's excrement" (LeMenager *Aesthetics* 73). It gushes from the earth voluminously, inspiring awe and ecstasy from onlookers, whether or not they have financial interests in the gushing well. Oil is extractive on two levels: it requires physical extraction from the ground as a nonrenewing resource produced by nature, and it requires value extraction from the laborers whose dangerous work supports the capitalist accumulation characteristic of big oil corporations (Hitchcock 85).

True, oil is a material commodity that we can identify outside of our skin, but in its harder-to-grasp state it suggests a cultural logic, a social convention that (when recognized) sparks moral debates about right and wrong behavior in economic and ecological spheres (Hitchcock 81). These layers of significance, economic, cultural, and personal, are formative in America's cultural evolution and self-definition (Hitchcock 82). Our reliance on oil is overtly economic, but it also involves buried psychological layers of compulsion and resistance to alternatives. Peter Hitchcock has argued that the saturation of oil in every aspect of our daily lives "paradoxically has placed a significant bar on its cultural representation ... everywhere and obvious, it must be opaque or otherwise fantastic" (81, 87). Since literature serves to represent and critique cultural hegemony, literary oil is a double-sided trope. It both celebrates the troubling enabler, as when Jack Kerouac's Sal and Dean roar through small Midwestern towns in a behemoth Cadillac limousine, and staunchly opposes its influence, as when Edward Abbey decries car-based tourism in the great western national parks. Petroleum is the star of a geopolitical parody: yee-hawing Texas wildcatters rub shoulders with slick self-pampering desert Sheiks skiing in air-conditioned malls and postcolonial African autocrats on oil thrones, clutching wads of dollars instead of scepters.

Since each liter contains about as much energy as a man laboring for five weeks, it is little wonder that when we buy this slave-in-a-bottle for $1 per liter, it quickly becomes our master (LeMenager *Aesthetics* 60). Americans are used to the highest levels of energy consumption in the world; in 1988, the United States had 5 percent of the world's citizens, but consumed 25 percent of the world's oil and released 22 percent of global carbon emissions—an energetic calculus that is equivalent to seventy-three healthy slaves laboring for each citizen (Nye 6). In Bob Johnson's trenchant analysis of modern oil ontology, the equation of mineral energy to slave labor is more than a striking analogy. The correlation between the petro-slave in the gas tank and the physical labor of our ancestors provides a narrative of the "dramatic dislocations that came with transitioning from a low-energy world of hooves, feet, and hands to a new and unfamiliar one wherein work was increasingly disembodied ... a sign that the most elemental relationships among energy, work, and the body were being recalibrated" (41–2). As the human body consumes energy far beyond the confines of the skin, cultural critics consider how the industrial body was redesigned around the phenomenal energies of ancient sunlight-turned-carbon and the technologies that allow us to control and deploy that fantastic power. The body's own matter is no longer built of annual sunlight, but of coal and oil, and this "embodiment of fossil fuels" correlates with the "disembodiment of labor" wherein an abstract thermodynamic calculus, actionable on much larger scales than humans and horses, substitutes for the somatic labor of a basic muscular economy (Johnson 43–5).

No wonder American culture has always resisted the Malthusian logic of ecological (and therefore economic) limits. Its 250-year history has been defined by using extraordinary stores of natural resources—first dense, vast forests and fertile soil, then coal, then oil—to defy Malthus's checks on population and quality of life. The American twentieth century was the most energetically rich in human history. Before the created crisis of the oil embargo in the 1970s, America never experienced the crushing limitations of energy that were quotidian in older, highly populated, deforested European locales (Johnson xix). Though American exceptionalism cites a Protestant work ethic that drove industrialization and productivity, a diverse and talented immigrant population, and a Christian faith-based sense of manifest destiny as features of a unique nation, the material conditions these Americans inherited were the truly exceptional factor. The United States had virgin forests *and* fertile land *and* coal *and* petroleum; infrastructure could be created ex nihilo and cities chartered without concern for preceding centuries of entrenched culture and design. In many ways America defined itself against the European model—of course in the sociopolitical sphere, but also in valences of planning and infrastructure, valuing dispersal and decentralization.

The upside's downside was the binge of energy that encouraged policy choices favoring consumption and rapid expansion over efficiency and wise growth. Part of the legacy of that fossil-fueled mania is that Americans now consume much more energy per capita than European counterparts, without gains in quality of life. Americans may even be said to suffer from their energy appetites, as the norm of the twentieth century became embodied in soulless suburbs, the car commute, sedentary lifestyles, and abundant, cheap, low quality, un-local food. This strange situation has come to seem normal. Paradoxically, America's energy riches have led to a cultural devaluation of that energy, as abundance and dependence cheapen our perception of the value of energy (Johnson xxi).

Oil's extraordinary chemical energy density encourages cultural excesses that mark the consumerist fetish of energy overabundance: yellow Hummers and the McDonald's "Dollar Menu." Shame and celebration are the inhale and exhale of a bipolar oil culture. The developed world has reached a point at which the obscene display of excess energy is met not only with grassroots opposition, but also with major subcultures favoring walk and bike communities, local and slow food, and an array of repurposed and recycled consumer goods. Petroleum is varied enough to support these trends, and we have not lost our muscles altogether. The sun, the wind, the rivers, the tides, and the earth's hot heart all still work without oil. But in the second decade of the twenty-first century, this is still a well-oiled world—designed, erected, and maintained by oil conglomerates—with several billion avid buyers.

America's history of energy use is phasic: biomass (wood) was the almost exclusive fuel source until the discovery of domestic coal in the mid-nineteenth century. America's industrialization after the Civil War was fueled by coal; Bob Johnson cites the year 1885 as a pivot point between reliance on biomass and fossil fuels (xviii). Once large oil fields were opened in Texas, Oklahoma, and California in the first years of the twentieth century, coal use plateaued and soon declined. The twentieth century would be defined by America's modern industrial economy designed to flow on petroleum, with roads replacing rails. By the end of the Second World War, oil fueled more energy demands than declining coal, and the trend continues to this day. The renewable energy market has made significant gains in today's economy, but it is still a small fraction that suffers in the competition with low oil and natural gas prices.

With the United States leading the world both in the history and pace of oil consumption, it makes sense that a lot of literary oil pumped through Yankee pens in the twentieth century. The national creed of liberty and self-reliance feasted on oil in the era of the wildcatter, the fabled individual roughneck who makes a fortune through his pluck and good luck. That individualistic enterprise soon translated to the oil corporation, a legal individual permitted to defend its interests as though it were a person.

The richest man in history, John D. Rockefeller, was buoyed to the top on a series of good-time gushers and merciless corporate predation. With its rich endemic supplies of oil, the United States showed the world how to extract and use oil to support sprawling industrial transportation—the national highway system. As we saw with coal, the nineteenth century established the world's first commuter suburbs along railway lines, but it was oil that enabled the United States to drive a paradigm shift in the geography of residence. As US domestic supplies have long been on the wane, the superpower flexed military muscles to claim oil supplies on other continents, beginning an era of corporate colonialism that has had particularly strong effects in Africa and the Middle East. If the first colonial era, managed from European metropoles, targeted locales rich in agricultural resources and cheap labor to supply raw materials to coal-driven manufactories, the second era of corporate colonialism began under American management with a clear target of mineral resources and a capitalist imperative to create market desires that oil would fulfill. The cultural Americanization of the world is a tale of a barrel of oil.

II Gusher gawkers

Compared to the dark, squalid horror of the coal mining disaster, the oil gusher is a disaster that excites and stimulates the onlooker, doubling as a figure of arousal and ejaculation (LeMenager *Living Oil* 93). It's the most explicit representation

Figure 6.1 Anonymous, *The First Oil District in Los Angeles, Toluca Street*, c.1895–1901. Image courtesy of Wikimedia Commons. Public domain.

of orgiastic power in oil's repertoire, a spectacle that, according to Michael Ziser, justifies "almost any degree of fetishization" (321). Upton Sinclair's industrial epic, *Oil!*, an urtext of any literary study of petroleum, ogles the gusher and its sublime destructiveness. Part bildungsroman, part declensionist oil narrative, part author's opinion column, the book traces the development of oil wells in southern California through the voice of young Bunny Ross, son of oil tycoon J. Arnold Ross. The novel explores capitalism versus socialism, the technical details of drilling and their environmental effects, the beautiful people of Hollywood and the poor ranchers of inland California, and, as the base of it all, the agony and ecstasy of a world blown open by petroleum power. Sinclair balances a boy's reverence for his oil magnate father with his awakening qualms about the social and ecological impacts of oil drilling. The novel does not address the larger issue of oil economics as part of America's imperial ambitions in the American century, but it does stay on-site to witness the extreme transformations oil produces wherever it is discovered (Hitchcock 90).

Bunny's father foreshadows the individualist baron that Ayn Rand would raise to worship status in the 1950s with her profit-driven heroes like Hank Rearden and John Galt. The fortunes of these two characters were built upon the railway, not oil, but Rand tolerates no tragedy in her polemical novels— her heroes win every time, after extended soliloquies on the virtues of free enterprise. Sinclair's J. Arnold Ross captures the spirit of the American twentieth century better than Rand does, with his story of hard work from modest birth, the death grip of the successful tycoon, and the tycoon's inevitable decline as the industry splinters due to troubles with labor and resources. Rand and Sinclair take perfectly inverse perspectives on the narrative of the esurient industrial baron. Sinclair composes the more balanced and textured account of the American story from a Socialist perspective, while Rand's simplistic no-apologies, no-prisoners heroism in which oligopoly is interpreted as free-market capitalism, has become gospel to some Libertarians. The point of unity between these diametric writers is their fascination with how the human desires for power and control trump the inherent human capacity to foster collective good and altruism. Rand seems not to believe in the latter, and pillories those who do. Sinclair believes in them despite all the evidence of selfishness he explores in nauseating detail in *Oil!* and *The Jungle*. In his naïve, childish thrall Bunny reflects, "Never since the world began had there been men of power equal to this. And Dad was one of them" (6). Bunny would also, in turn, be anointed by oil, and then traumatized by the conditions and costs of its production.

Sinclair begins with an ode to the open road, with events set in "automotive time" and distance felt as an abstraction (A. Wilson 34):

> Any boy will tell you that this is glorious. Whoopee! You bet! Sailing along up there close to the clouds, with an engine full of power, magically harnessed, subject to the faintest pressure from the ball of your foot. The power of ninety horses—think of that! Suppose you had had ninety horses out there in front of you, forty-five pairs in a long line, galloping around the side of a mountain, wouldn't that make your pulses jump? And this magic ribbon of concrete laid out for you, winding here and there, feeling its way upward with hardly a variation of grade, taking off the shoulder of a mountain, cutting straight through the apex of another, diving into the black belly of a third; twisting, turning, tilting inward on the outside curves, tilting outward on the inside curves, so that you were always balanced, always safe—and with a white-painted line marking the centre, so that you always knew exactly where you had a right to be—what magic had done all this?
>
> Dad had explained it—money had done it. (5–6)

Soaked in black gold, 1920s America found itself rich, in geology and ambition. Bunny's thrill comes from his translation of biomass power—the imagined train of horses that would impress even Santa Claus—into fossil power yoked to a mechanical drive train and enabled by the roads that pave a particularly American experience. In his glee, Bunny sails on ribbons and gallops over the bent backs of mountains. But despite all the potency of energy, the well-bred scion sits next to his weighty father, "silent and dignified—because that was Dad's way, and Dad's way constituted the ethics of motoring" (1).

Bunny's is the first generation to experience the annihilation of distance, of space and place, by petroleum. The Victorians felt the awful fossil power of coal, deployed by locomotives and steamships, but Bunny's experience of oil is different. His is an individual power, individually experienced, and available at the individual's demand: a new iteration of fossil superpower. The landscape they hurtle over, privatized by their automobile, signifies only a blank canvas streaked by oil. Mountains are no longer obstructions. The magic ribbon cares about one place: the individual's destination.

From the heady mountain heights, Bunny and Dad descend into a new kind of snarl invented in long-suffering Los Angeles:

> By dint of constant pushing and passing every other car, Dad had got on his schedule again. They skirted the city, avoiding the traffic crowds in its centre, and presently came a sign: "Beach City Boulevard." It was a wide asphalt road, with thousands of speeding cars, and more subdivisions and suburban

home-sites, with endless ingenious advertisements designed to catch the fancy
of the motorist, and cause him to put on brakes. (21)

Sinclair sniffs out oil's cultural geography a few decades before Eisenhower
would pave the way with the interstate highway system: viral suburbia and inane
roadside attractions. People in a petrotopia live by the sparkle promise of the
billboard and subdivision. *Have everything now*, chirps the commercial bird.
This kind of utopian vision of the automobile has a despotic underside: if utopias
reflect spatial ordering imposed by industries (automakers, developers, oil
companies) to serve their own need for profit, petrotopias signify economic and
political will rather than an ideal state of civilization (LeMenager *Living Oil* 74).
When everyone is a consumer within a petrotopia, new problems emerge, and
the saleable utopian image becomes tread-marked with the realities of its false
design: not to make people happy, but to make them continually less than happy,
that is, continually consuming. The car-serving strip is an endless regression of
franchise businesses, standardized images and logos, parking lots, and flattened,
simplified landscapes (A. Wilson 33).

These are the major challenges of American civil infrastructure into the
twenty-first century, particularly the car's blending of city economy with
country leisure, adulterating both in the process. Within the first few pages of
the book, Sinclair points out a startling paradox of the automobile landscape: it
erases the spaces with inherent natural value while vulgarly highlighting places
of vacuous commerce. The car elides and flattens deep spaces, where mountains
and rivers dwell; it plumps up shallow spaces hocking recursive baubles. It could
be argued that the automobile actually opens wild lands to access and therefore
appreciation, but literary history is very skeptical of this road-trip mentality, as
we will see with Edward Abbey.

Beyond question, the network of roads and interstates built to accommodate
the automobile have enabled the existence of large-scale commerce on the
American model, and the petrotopia is one of our more enduring exports.
Homogenized goods, desires channeled through advertising, indicate the next
paradox of petroculture: the individual freedom premised by the automobile
actually cultivates mass normative behavior. Petropeople are forever stuck
together in the traffic jam at the mall. Oil catalyzes a revolution of space and
commerce in the modern state, and then it shackles our ability to imagine
beyond its own enabling logic of mobility and accumulation (Hitchcock 91).

One little traffic jam is no match for Bunny's oil ardor, however; it merely
enriches the spectacle unfolding outside the windshield. He is soon treated to

the seminal iconic image of the oil era: the gusher. Gushers became important symbols after the discovery of oil at Spindletop, near Beaumont Texas, on January 10, 1901. That first Texas gusher roared unstaunched for nine days before the prospectors could cap it. It brought tens of thousands of wildcatters to the pinewoods of east Texas, sniffing the ground for black gold. A week later, on January 17, the *New York Times* reported "the greatest oil strike in the history of that industry ... oil shoots out of a six inch pipe to the height of seventy-five feet ... nearly 50,000 barrels of oil have flowed into the ravines nearby. These ravines are being dammed as rapidly as possible, but many thousand barrels of oil have already gone to waste." The black muck was enriching the ground it saturated; in those first weeks after the Spindletop discovery, "fabulous offers" on adjacent land leases poured in, and many claims were only as large as a derrick foundation. Depression follows close upon manic exuberance: the oil field was tapped out within two years.

Sinclair's character J. Arnold Ross rushed through mountains and suburbs to reach his claims at Prospect Hill, where a gusher signaled the greatest new oil discovery in southern California:

> The inside of the earth seemed to burst out through that hole; a roaring and rushing, as Niagara, and a black column shot up into the air, two hundred feet, two hundred and fifty—no one could say for sure—and came thundering down to earth as a mass of thick, black, slimy, slippery fluid. It hurled tools and other heavy objects this way and that, so the men had to run for their lives. It filled the sump-hole, and poured over, like a sauce-pan boiling too fast, and went streaming down the hillside. Carried by the wind, a curtain of black mist, it sprayed the Culver homestead, turning it black, and sending the women of the household flying across the cabbage-fields. Afterwards it was told with Homeric laughter how these women had been heard to lament the destruction of their clothing and their window-curtains by this million-dollar flood of "black gold"! (25)

Silly women worry about small domestic things. As they ogle the deluge unleashed by earth's pressing crust, the mighty oil barons, gods on Olympus, laugh at the melee. The gusher is an awesome force of nature, an ejaculation of earth fluid that is too sublimely gross not to watch, and too valuable not to respect and envy the person with the drilling rights. The Culver homestead, once a place of clean curtains and cabbages, has in an instant become a soiled relic on "porcupine hill" where dozens of derricks will soon bristle. The hill has changed from a place of surface solar agriculture to a subterranean fossil pool, a shift that is at once spatial, temporal, cultural, ecological, and economic. It will never return to its earlier condition. The suburbs will sprawl, the fossil pool will be drained, and the solar pastoral of cabbages will be trampled.

Sinclair does not pause to eulogize the downtrodden pastoral past, however. The tragicomedy of the gusher has the charisma to dominate the stage:

> Meantime the workmen were toiling like mad to stop the flow of the well; they staggered here and there, half blinded by the black spray—and with no place to brace themselves, nothing they could hold onto, because everything was greased, streaming with grease. You worked in darkness, groping about, with nothing but the roar of the monster, his blows upon your body, his spitting in your face, to tell you where he was. You worked at high tension, for there were bonuses offered—fifty dollars for each man if you stopped the flow before midnight, a hundred dollars if you stopped it before ten o'clock. No one could figure how much wealth that monster was wasting, but it must be thousands of dollars every minute. Mr. Culver himself pitched in to help, and in his reckless efforts lost both of his ear-drums. "Tried to stop the flow with his head," said a workman, unsympathetically. In addition the owner discovered, in the course of ensuing weeks, that he had accumulated a total of forty-two suits for damages to houses, clothing, chickens, goats, cows, cabbages, sugar-beets, and automobiles which had skidded into ditches on too well-greased roads. (26)

Monsters are hard to love, but the avarice of these grease monkeys overcomes the obscenities of their work, and a toxic passion takes a toll on the devoted Culver, who is deafened by the roar of money. Litigiousness ensues on these uneasily shared pastures, where Culver's wealth and carelessness present an opportunity for others to cash in. Chickens and sugar beets are suffocated in oil and exchanged for oil money. The homesteaders' collective survival evolves into an economy of individual competition. The whole landscape is now oil, along with the frenzy and animosity that accompany fast riches. Behind oil's obvious representation of new wealth lingers a set of emotional histories that are scripted onto the oil: hope, fear, anxiety, and aspiration (LeMenager *Living Oil* 93). Stephanie LeMenager's final answer to her own query, "Why is oil so bad?" relates to the spectacle of oil best symbolized by the gusher: "We gather to watch" (*Living Oil* 101). What we watch begins with the gusher but flows out to the whole culture of voyeurism enabled by the anonymous, independent movement of the car and the spectacles of an energy-saturated culture.

Sinclair's long novel follows Bunny from his naïve adoration of the spectacle and his doughty father into his growing awareness of the victims of oil culture—the nonunionized workers, the hapless landowners hoodwinked by barons, and the oil-giving land itself, which is defiled not only by the primary pollution of the gusher, but on a much larger scale the land is transfigured by the actions made possible and profitable by the oil: movement, settlement, and development. When Bunny and his father go oil prospecting under the false pretenses of an innocent quail-hunting trip, Sinclair brings nostalgia for lost preindustrial times directly into the maniacal path of the future: their actual oil-going motivations directly result in the death of the natural habitat they pretend to enjoy. Once Ross establishes his claim by outmaneuvering the landowners, the derricks, the gusher, and the black blight tumble in an inevitable narrative series. Tragedy and injury without humor are Bunny's impressions of the gusher the second time around. Sinclair pushes Bunny's witness from admiration to awe to sickened disaffection. An activist is born.

Sinclair's fiction is built of the elements of real history. Once the sheer use value of gasoline had established thirsty markets and spawned the American romance with the automobile, petroleum companies deployed the Texas wildcatter as an embodiment of the risk-hungry, wily, egotistical, free-enterprising, self-made man so readily worshipped in public discourse. This characterization gelled with gasoline ads that commonly depicted juiced-up consumers in thrall of their powerful cars, the new proselytes of the open road. The February 13, 1950, cover of *Time* magazine shows wildcatter Glenn McCarthy with a frontier suntan and tousled hair glancing back at a drilling tower ("Texas: King of the Wildcatters"). "Diamond Glenn" is wearing a Houston-brass pinstripe suit, and the anthropomorphic drilling tower sports a Stetson, cowboy boots, and a pair of bulging biceps. The derricks and the gusher are Glenn's sidekicks in the cartoony popular culture of the oil cowboy. Who needs cows, quails, or a ranch when there are gushers ready to blow?

McCarthy comes across as a wildcatter for the corporate postwar twentieth century: dressed like a businessman but harboring the adventurous soul of a cowboy in the old west. The *Time* story begins: "If it were possible to cap the human ego like a gas well, and to pipe off its more volatile byproducts as fuel, Houston's multi-millionaire wildcatter Glenn McCarthy could heat a city the size of Omaha with no help at all. Whether he would allow his rampant psyche to be dedicated completely to so prosaic a project, however, is doubtful—several million cubic feet would undoubtedly be diverted to a McCarthy Memorial Beacon which would nightly cast its glare as far west as El Paso" (1). So popular was his charm-smacking, egotistical persona that he was replicated in Edna Ferber's 1952 novel *Giant*, adapted into the 1956 film. The appeal of this iconic wildcatter figure has not diminished in the intervening half-century: a version of the "Diamond Glenn" persona became a two-term American president in the twenty-first century. Though George W. Bush never soiled his boots at the derricks, his policies were rote recitations of big oil's agenda. The continuum between the scrabbly wildcatter, the glitzy Texas powerbroker, and commander-in-chief falls within a surprisingly short arc of history, and oil washes off the hand more easily than off the land. In each of these personae is a narrative that tends to credit audacity and self-confidence as pure virtues, to appraise elements of luck and resource availability as earned opportunities, and to erase what is defaced and destroyed.

III Road trip!

The single most potent symbol used in the rhetoric of American freedom is the open road. Though most Americans spend more time in local traffic running errands, the frontier-piercing highway continues to work powerfully on the psyche. Equally powerful emotions of frenzy and animosity weigh heavily in American driver's seats as the open road turns to the exurban imbroglio. These petro-propelled emotions are a psychological correlative to the fuel itself, the fantastically powerful fluid that contains, in a single liter, the same energy that a man would need five weeks of hard labor to exert. With such power at our fingertips and toes, after sitting behind the wheel, it is not surprising that petroculture quickly adopted ideals of speed and power in its machines.

The classic American cultural space, the open road, became a national policy and Americans invested in it heavily. Major grants for the construction of highways in 1947 and 1957 laid the foundation of 79,500 miles of pavement (Gill 34). Suburb-hater James Howard Kunstler notes that one major developer, Robert Moses, was responsible for converting rural Long Island's 1,400 square miles into "a parking lot" by the time he retired in 1968 (Gill 34). The interstate system, the legacy of President Eisenhower, is the most celebrated investment, but the entire twentieth century offers a similar history of conquering space and time with convenience and speed, and suppressing wildness with a tarmac of interconnected roads over the land. This highway lattice represents the creation of a new dimension of being—one in which origin and destination are fused, not by intermediary prairies or mountains that actually exist, but instead by a quintessential petro-cognitive dimension that feeds on speed, consumer convenience, and self-absorption.

As David Nye has memorably argued, technological momentum does equate to determinism: our world is designed around opportunities and choices, and the American love affair with the automobile was never a compulsory relationship. America had mass transit systems in place in the early twentieth century, which we chose to dismantle and neglect in preference for the power, convenience, and freedom car producers promised. In Nye's words, "The use of the car in America expressed cultural values, not a technological imperative. The automobile ministered to a pre-existing penchant for mobility. It did not cause suburbanization or nomadism; Americans used it to express these choices" (177). Once the choice was made, however, once the interstate highway system existed, the system became more rigid, less open to choice, more devoted to the automobile as a way to recoup costs for road construction and maintenance. All the other effects of the personal car—low-density suburban development, roadside culture, fragmented natural habitats—flowed downstream of that choice point.

The American road trip is only superficially about the places where travelers pause before spinning down the line; roadside attractions, natural and cultural, are constructs of the highway ideology of open access and marketed interest. To stop at South of the Border (SOB—a chain of restaurants off I-95 in the southern United States) is to indulge in an experience of cheap representation. For hundreds of miles in either direction, "Pedro," the stereotypical Latino mascot, shrills about the thrills of SOB, informing travelers that their experience of this petroculture is essential to the patriotic car-borne experience. "You never sausage a place!" Pedro pipes from a billboard studded with an enormous plaster wiener. True, Pedro: SOB does attain that unique American roadside quality of openly, knowingly cultivating kitsch for its own appeal, rather than being a naive mistake of cheap commerce gone too far. SOB and its cohort of show-me stops are the playground of a now-global culture of tacky, instantly gratifying, endlessly available baubles, many of them actually made of petroleum. It's a gas.

The American road trip is fundamentally about discovering one's self in relation to nothing in particular—or at least, nothing lasting. The road traveler's shadow pitches over a hundred different scenes, a mile or more per minute, traversed with such ease and speed that the places are little more than sketches of themselves, forgotten as soon as the next frame comes into view. What one gets on a road trip is time to contemplate origins and destinations, past and future, to the amusing scenic musak of the fleeting present. The only vestige the visitor leaves behind is the imprint of her credit card. To consider how literature has entertained this transient state of the present tense, two novels in particular will serve: *On the Road* by Jack Kerouac (one frenzied passage in particular), and *Desert Solitaire* by Edward Abbey (one angry park ranger–author in particular). Frenzy and animosity, travel and traffic, road rash and road rage: petroculture is our pet Cerberus. These texts show a persistent American confusion that equates driving with being alive, with being powerfully embodied by the prosthetic automobile (LeMenager *Living Oil* 80).

Novels have the power to represent entire generations, especially if the book records a flashy rejection of the previous generation by way of wandering, sex, and drugs. Consider the perennial popularity of *The Great Gatsby*, leading candidate for the elusive "great American novel," and itself a perspective on petrotopianism. George Wilson's filling station is not just a roadside attraction conveniently midway between the Eggs and Manhattan. It is the nucleus of the novel's action: the place where commerce meets God on the billboard, the place where the classes intermingle and abuse one another, the place of American small business meant to be ennobling, but which finally serves as a stage for infidelity and drunken manslaughter. The manic ecstasy and rage of petroculture collide, appropriately, at Wilson's garage.

Jack Kerouac's autobiographical *On the Road* examines the petroculture of nowhere. Kerouac was king of the Beats, the handsome, daring, reckless, rough, football-playing hero who spat in the face of 1950s conformity. At one ennui-filled station of life, his autobiographical characters Sal Paradise and Dean Moriarty acquire a 1947 Cadillac limousine. This car was six feet tall, nineteen feet long, seven feet wide, weighed over five thousand pounds, and was propelled by a 5.7 liter V-8 engine of unknown (and unspeakable) mileage per gallon. What a car for a romp from Denver to Chicago, blitzing across those monotonous plains and sleepy farm towns! This kind of car makes no apologies, it makes impacts. As Patricia Yaeger has noted: "Kerouac's characters are gasaholics. Oil dependency created their world; each city, suburb, truck stop, and bite of pie depends on Standard Oil" (306).

Sal's account of the trip demonstrates how this oil-driven existence is at least as much cognitive and ontological as it is physical:

> In no time at all we were back on the main highway and that night I saw the entire state of Nebraska unroll before my eyes. A hundred and ten miles an hour straight through, an arrow road, sleeping towns, no traffic, and the Union Pacific streamliner falling behind us in the moonlight. I wasn't frightened at all that night; it was perfectly legitimate to go 110 and talk and have all the Nebraska towns ... unreel with dreamlike rapidity as we roared ahead and talked. It was

a magnificent car; it could hold the road like a boat holds on water. Gradual curves were its singing ease.

"You and I, Sal, we'd dig the whole world with a car like this because, man, the road must eventually lead to the whole world. Ain't nowhere else it can go—right?" (230–1)

The all-night drive is a visual illusion: Nebraska seems to be unrolling from Sal and Dean's future straight into their past, with each foot of ground occupying, at 110 miles per hour, only 0.0062 seconds of their lives. A mile that would take twenty minutes to walk passes by every 32.7 seconds. What would Ruskin say? Time-space is contracted from our evolutionary standard of walking down to about one-fortieth of its original size. Those Great Plains seem miniscule on petroleum. Not only has the world shrunk, but it has also abstracted, transformed into a surreal picture show in which the old-fashioned locomotive falls behind the car's pace under the light of the vanishing moon. In his dream reverie, Dean, at the wheel, imagines himself much larger because the world appears so much smaller; it is barely memorable because it has never really been perceived. Nebraska is a flat, fast, darkness.

Sal and Dean have an idea of space, the impression of time, which the gods might share as they gaze down from Olympus. To quote Tennyson, the passengers are drugged and lulled by passivity like "The Lotos-Eaters," who recline in their seats "like Gods together, careless of mankind. / For they lie beside their nectar, and the bolts are hurl'd / Far below them in the valleys" (ll. 155–7). The nectar is petroleum, the bolts are eight valves punching power to the wheels, and Dean, complete with his frenzy of power and caprice, is Zeus.

The liberty from space and time made possible by fossil fuels carries with it a sensual ecstasy that frequently blends with manic madness: it is a road fiction that imagines how it feels to live, and to die, by automobile (LeMenager *Living Oil* 91). Dean and Sal use the car as a way to chase *thanatos*, the death drive that is thrilling because of its inherent risks. The physical experience of raw power and momentum far beyond the body's limits had previously belonged to a world of dream and fantasy, but the carbon-borne experience grants us a mighty prosthetic (Johnson 52). From Charles Lindbergh's wings high above the Atlantic to Kerouac's maniac wheels belittling the Great Plains, the modern, oil-borne body is a cyborg that trains muscles to maintain machines. Our bodies no longer end at our toes and fingers, but somewhere further—the wheel, the pedal, the carbon-made pavement. These fantasies even fit into postapocalyptic visions of future scarcity with films like *Mad Max: Fury Road* (2015), in which a feature-length road war unfolds as a dialectic of aggression and fuel consumption.

Genders are neutralized: the women rev their engines with as much fury, and often greater success, than their male counterparts, so this very masculine film has been hailed as a feminist triumph. But it can only be considered so within the petro-fed zeitgeist of the culture from which it emerged: the early twenty-first century, a full tank of angst and exuberance.

The open road is the surreal ideal, and such flights of altered consciousness, the psychic and erotic highs, do not last. Space and time eventually run out and wrestle the flesh body to the earth. The next day, the rest of humanity returns, and Dean's night of petro-ecstasy turns into zombie-like withdrawal. Sal recalls:

> I went to sleep and woke up to the dry, hot atmosphere of July Sunday morning in Iowa, and still Dean was driving and had not slackened his speed; he took the curvy corndales of Iowa at a minimum of eighty and the straightaway 110 as usual, unless both-ways traffic forced him to fall in line at a crawling and miserable sixty. When there was a chance he shot ahead and passed cars by the half-dozen and left them behind in a cloud of dust.
>
> Great horrors that we were going to crash this very morning took hold of me and I got down on the floor and closed my eyes and tried to sleep. As a seaman I used to think of the waves rushing beneath the shell of the ship and the bottomless depths thereunder—now I could feel the road some twenty inches beneath me, unfurling and flying and hissing at incredible speeds across the groaning continent with that mad Ahab at the wheel. When I closed my eyes all I could see was the road unwinding into me. When I opened them I saw flashing shadows of trees vibrating on the floor of the car. There was no escaping it. I resigned myself to all. And still Dean drove, he had no thought of sleeping till we got to Chicago. (234–5)

The too-fast-too-furious scene goes from a literal description of Dean's obstacles to Sal's figurative horrors at the terrible forces at play. Not only the forces of physics—the multi-ton car and its velocity, F=ma—but also the force of Dean's compulsive personality. Dean is the deranged captain, bent on annihilating everything between himself and the white whale of Chicago, and killing Chicago once they find it. The weapon is the road itself, "unwinding into" Sal like a lance, threatening to pierce his prone body at every turn. They do wreck, they do have to stop, but those pauses serve as way stations where they assess how much road they have so quickly conquered. The car roars, the continent groans, and the battle hastens toward its conclusion at Chicago, with a tamed half-continent left in their dust: 1,180 miles in seventeen hours, "a kind of crazy record" (238). After this smug summation, Sal and Dean descend back into beat-down status. The car reflects their condition as raggedy outcasts, just a few hours after anointing them gas-guzzling gods of the road:

> At intermissions we rushed out in the Cadillac and tried to pick up girls all up and down Chicago. They were frightened of our big, scarred, prophetic car. In his mad frenzy Dean backed up smack on hydrants and tittered maniacally. By nine o'clock the car was an utter wreck; the brakes weren't working any more; the fenders were stove in; the rods were rattling. Dean couldn't stop at red lights, it kept kicking convulsively over the roadway. It had paid the price of the night. It was a muddy boot and no longer a shiny limousine. (238)

This account of the ferocious arc of fuel consumption is among the more famous car trips in literature, a classic vignette of the exhilaration and horror enabled by oil. In the mid-twentieth century there were many ambivalent literary accounts of petroculture, though none with as much sophomoric verve.

A contrasting impression of vast Nebraska comes from another novel completed in 1951, Patricia Highsmith's *The Price of Salt* (or, *Carol*). The reader rides in the mind of a young woman named Therese, who is on the lam with her lover Carol. On the level of oil ontology, Highsmith's novel can be seen as part of the "tilt in sexual mores that the car allowed, especially to young women" (Ziser 321). The women evade a detective who gathers evidence against Carol's moral competency to retain custody of her daughter. The entire novel is a psychological tornado, with Therese overwhelmed by the euphoria and fear of a first, and forbidden, love. This passage comes just after Carol was ticketed for speeding, and Therese's infatuated mind reads significance in the nuances of the moving landscape:

> Therese offered to drive, but Carol said she wanted to. And the flat Nebraska prairie spread out before them, yellow with wheat stubble, brown-splotched with bare earth and stone, deceptively warm looking in the white winter sun. Because they went a little slower now, Therese had a panicky sensation of not moving at all, as if the earth drifted under them and they stood still.... . She watched the land and sky for the meaningless events that her mind insisted on attaching significance to, the buzzard that banked slowly in the sky, the direction of a tangle of weeds that bounced over a rutted field. (192)

Less speed causes adrenaline withdrawal, but it also heightens fear and paranoia in the fleeing subject, who feels her pursuer gaining ground. Therese, car-imprisoned, occupies herself with the only pastime available: to watch the scenery. Sal and Dean did not look at scenery, they felt the velocity. Therese, on the other hand, contemplates it so deeply that it comes alive with signification. Is the buzzard a spy, or herself caught in circles of thought? Does the tumbleweed have a destination, or is the wind an irresistible but aimless force that blows us around? Releasing the gas pedal, at least to sub-manic speeds, reverses the petro-psychology from Dean's inward-to-outward, where the landscape was a screening of his fears, to Therese's outward-to-inward, where the landscape offers features that yield introspection. She arranges her mind around the natural composition. Of course, the women are not sitting still, they are moving at about fifty miles per hour. But in this 1950s pre-interstate oil ontology (I-80 did not cross Nebraska until 1957), easing along the smaller, rougher roads allows the landscape to emerge in relief, as an assembly of living, dynamic inhabitants that might come in conversation with the rolling subject, rather than just crossed (out) by the speed-blind. In this case, the conversation is depressing. If in the 1950s the speeding car is a metaphor for progress, here Highsmith accomplishes an intriguing reversal of this highway ideology by trapping her characters in existential limbo (A. Wilson 34). They make use of the same infrastructure—car and highway—as other Americans, but their gender and sexual orientation convert mainstream freedom and progress into its opposites: pursuit and imprisonment.

A few days later, Therese finds herself with the car but without Carol, who has returned to New York to triage her child-custody standing. Again, speed emerges as psychological:

> It was a fine day, cold and almost windless, bright with sun. She could take the car and drive out somewhere. She had not used the car for three days. Suddenly she realized she did not want to use it. The day she had taken it out and driven in up to ninety on the straight road to Dell Rapids, exultant after a letter from Carol, seemed very long ago. (215)

If Therese could not feel Carol's touch, she could substitute speed for the thrill, and the cursive word for the voice. But laid low by neglect, Therese cannot replace intimacy with speed, and the failure of petro-ecstasy to stand in for real life is poignantly aggrieving. The film version of *Carol* (2015) depicts a much shorter journey from New York to Iowa, the emotional arc of road tripping is shown using physical cues: the happy westward journey is high on champagne, giggles, and gazes; the sad eastward return is marked by Therese vomiting roadside and sleeping fetally in the back seat. The oil, the car, the road, and the liberty are all still there. Carol is not.

Both Kerouac's and Highsmith's couples illustrate how the road served countercultures of the 1950s, as well as embodying the postwar mainstream appetite for liberty and consumerism. The beatniks Dean and Sal, and the lesbians Carol and Therese, both find spaces in which to think and feel as they naturally are, rather than as society expects them to be. Nebraska's open road is a canvas of self-realization at the same time as it is a delusion. The road ends: Dean must return the vandalized Cadillac to its owner in Chicago; Carol must fly to New York to face her judges. Their companions Sal and Therese suffer the withdrawal of addicts, though the nature of the addiction is complex. The oily human condition injects physical velocity into emotional love. The industrial-era lover so often conflates emotion with motion that it has become cliché. Leaving on a jet plane. Hit the road, Jack. Baby, you can drive my car. Life in the fast lane. Life is a highway. Radar love. Get outta my dreams, get into my car! Shut up and drive.

When the oil runs out, will we resume writing ballads to our mistress's eyebrow?

The American car ruled the interstate, but it also infiltrated the scenic byways of textured land—from the intimate folds of Appalachian valleys, to the quilted plains, to the miles-high mountains, to the coasts, rocky and sugar sand. Prospectors graded and paved centuries-old wagon and hoof roads, and surveyed and laid thousands more ex nihilo. These byways were designated spots of beauty to be collectively enjoyed and tax supported; their paths changed forever the character of the map itself and the places the roads opened to traffic. Most essentially, the things located along good roads became defined by their convenient accessibility.

Places far from paved roads remained wilderness, compared to those places exposed to the parade of pleasure tourists seeking nature on their own, industrial–consumerist terms. The interpolated frontier was opened further with each two-lane road that tunneled into obscure nooks and made them marketed destinations. For nature-cultural theorist Alexander Wilson, the Blue Ridge Parkway represents the acme of landscape management, a ribbon of industrial-era construction that displays utter control of the field of vision, separates productive from nonproductive landscapes so the visitor may enjoy a "pure" experience, and produces an impression of nature itself as a series of magnificent vistas organized around a tourist calendar: vernal rhododendrons and laurels, autumn colors, and winter wonderlands (A. Wilson 37).

Ralph Seager's 1959 poem "Owed to the Country Road" is an elegy for the dirt-top tracks lost to the era of paving. The aesthetic of the road shifted from the organic meander of old dirt tracks laid by centuries of feet and wagon wheels toward an engineered efficiency based upon least distances. Seager's first accounting of the loss is based on color and light. He misses the "shadowed breeze" on his skin, his feet on the "quiet, cool, receptive clay," and the woods "crowded in / to watch with green eyes what was going by" (ll. 3, 7, 4–5). Embedded in the animate woods, the old country road touched the feet and hooves of travelers. The winding route allowed for surprise, the "unexpected view / of doe and fawn, or Paisley pheasant's brood / to grace and color rural solitude" (ll. 12–14). These intimate encounters between passing individuals were being replaced by bright impersonal vistas arranged by planners to make the road scenic, attractions for car-borne vacationers who perceive only "road and open sky" (l. 6).

Seager notes the violence brought upon the land to convert it into a modern thoroughfare serviceable to long-distance travelers. One packed iambic pentameter line identifies the perpetrators: "Bulldozers gouge and steel blades cut straight through," and the final effect of the development is one of death by engineering: "I see my hills knocked down, my valleys spanned / Where roads once fitted kindly to the land" (ll. 11, 15–16). Seager's poem updates the Romantic traditions of Wordsworth and Ruskin, who protested the incursion of rails into places that had been preciously rough; not wildernesses uninhabited by humans, but landscapes in which a long-evolved synergy between nature and culture had engendered unique locales. A large part of this protest tradition is the vulgar homogenization of space and the straightening of organic form into mechanical precision.

Edward Abbey's *Desert Solitaire* contains a polemic against inroads into public lands, particularly the national parks that contain the most exceptional natural forms. Abbey is an environmentalist version of a beatnik, living outside of a mid-century American culture and peering in, with fear and disdain, at its values and waste. His cantankerous voice serves as counterpoint to the Smokey Bear jollity of the self-promoting Park Service. As a ranger at Arches National Monument in the 1960s, Abbey witnessed the survey crew arrange a hardtop road that would go deep into the park, within strolling range of many of the reclusive arches: an arroyo of tar put down to attract the masses in touring cars and campers.

The survey crew's leader, both a "soft-spoken civil engineer" and a "very dangerous man," tries to buoy Abbey's spirits with the promise that his lonely dwelling in the wilderness—populated mostly by snakes, ants, and "Giant Hairy Desert Scorpions"—will soon bustle with the "indolent millions," a camper civilization in the wilderness enabled by cheap petroleum (54). The engineers see access and progress; Abbey envisions the campsite as "a suburban village: elaborate housetrailers of quilted aluminum crown upon gigantic camper-trucks of Fiberglass and molded plastic; through their windows you will see the blue glow of television and hear the studio laughter of Los Angeles" (44). The Park Service strives for exponentially higher visitation rates to Arches National Monument, which Abbey tweaks into "National Money-mint." Abbey sees the industrial tourist as a doltish victim of a "syphilization" that treats nature according to the needs of industrial logic. The access promised by the tarred roads and electric campgrounds becomes an infrastructure demanded by the machines, whose servants the people have become. Industrial logic inherently pursues saleable goods, and since nature itself, in the form of the Arches, cannot be sold beyond the entry fee, it follows that their appeal as wonders of natural history and ferrous-red beauty can be fed into the machine to derive saleable stuff: hotels, restaurants, refrigerator magnets, Smokey the Bear dolls.

Two centuries earlier, picturesque tourism on the backs of carriage horses began to establish the kitsch of a shoddy nature culture. The Claude-glasses, convex brown-tinted mirrors that allowed tourist to "frame" scenes that they would then sketch, flew off the shelves. In sketch books were scribbled many a masterpiece meant to adorn a lady's drawing room. English tourists in the Lake District could follow the series of stage-set roadside inns and expect ample food and shelter from the established preindustrial economy of the area. The difference between Wordsworth's tourist-friendly Lake District and the (American) National Park Service's improved campground in Arches is that petroleum makes the latter place survivable by a massive number of Great Indoorsmen. Without petroleum, there is no road, no water, no food, no electricity, and no television, and therefore nothing to sell. With it, the suburbanite's living room rolls with him into the wilderness, as does his wallet. Abbey writes, "So long as they are unwilling to crawl out of their cars they will not discover the treasures of the national parks and will never escape the stress and turmoil of those urban-suburban complexes which they had hoped, presumably, to leave behind for a while" (51). For Abbey, true appreciation of natural wealth is closely allied with the muscle power needed to experience it. In lieu of roads serving cars, he advocates trails serving hikers, horseback riders, and bicycles.

His solution has the added benefit of reversing the ontological shrinking of the world by fossil fuels. Where coal and petroleum have made the world smaller, closer to the roadside, and easier to swiftly bypass, flipping that history back a page to biomass fuel will have the enabling effect of expanding nature again. "Distance and space are functions of speed and time. Without expending a single dollar from the United States Treasury we could, if we wanted to, multiply the area of our national parks tenfold or a hundredfold—simply by banning the private automobile" (55). Naturalists can experience this reinfusion of distance into space by hiking into the secluded backcountry, which generally has no-engine and no-trace regulations.

In the 1960s, an era of cheap oil-driven consumerism, Abbey warns about the increasing physical conflation of humans with their machines, and sees the national parks as a final frontier against *Homo campervans*. In his polemic on industrial tourism, the ranger-ecoterrorist's final act is to tear out the survey team's flags by moonlight to erase the sketch of a road for just a bit longer. Fifty years later, on the other side of that incursion, many of the car-overrun national parks, including Yosemite and Zion, have recuperated peace with shuttle bus networks. Abbey's radical act has been incorporated into the progressive mainstream.

Let's face it: a gas-powered machine makes work easier. We work awfully hard trying to figure out how to save ourselves work. Gas can make extreme conditions survivable. For example, most people could not trek thirty miles across the searing Utah desert as the sun spans east to west. This does not deter the judgmental Abbey, who goes so far as to call the handicapped and children pro-automobile "pressure groups" whose needs are best ignored. Abbey's writing is as fraught and contradictory as it is saucily entertaining. While he decries roads through Arches in *Desert Solitaire*, he depicts his hero George Hayduke, one of *The Monkey Wrench Gang*, littering that same desert with beer cans tossed from his roaring truck. No-apologies irreverence is part of Abbey's misanthrope shtick, and it plays well with libertarians of many stripes.

Abbey's bile overflows in his highly influential novel *The Monkey Wrench Gang*, a thought experiment on the potentials of ecoterrorism to liberate the American West from industrial development—or, in his words, to "sabotage the world planetary maggot-machine" (151). The book became a founding text for fringe environmental groups like Earth First! who use sabotage as a weapon against development. Most often the gang's victims are dieselivores: trucks, tractors, bulldozers. These dumb machines are the prime movers that transmogrify the desert landscape from an unpeopled wilderness into what it is today: an electrified, irrigated sun strip, sprawling a thousand miles from Texas to California. The cities of Albuquerque, Phoenix, Las Vegas, San Diego, and Los Angeles, in particular, have benefited from twentieth-century projects that made mass habitation of these intensely hot, dry places possible. Petro-powered tools carved out and laid down the highways, the bridges, and the dams.

The ecoterrorist gang surveys their quarry: "Down in the center of the wash below the ridge the scrapers, the earthmovers and the dump trucks with eighty-ton beds unloaded their loads, building up the fill as the machines beyond were deepening the cut. Cut and fill, cut and fill, all afternoon the work went on. The object in mind was a modern high-speed highway for the convenience of the trucking industry, with grades no greater than 8 percent" (80). He sarcastically surmises that the engineers' ultimate goal is "a model of perfect sphericity, the planet Earth with all irregularities removed, highways merely painted on a surface smooth as glass" (80). Abbey has no qualms about depicting sabotage as a legitimate moral strategy against the oily invasion of the America's remote canyon country. In fact, he has fun with it. Since it is amusing to watch inanimate objects blow up, I will simply sample a few of the scenes in which the monkey wrenchers torch expensive equipment made possible by petroleum.

First, George Hayduke drives a bulldozer into Lake Powell, Glen Canyon Dam's impoundment:

> The thunder of the impact resounded from the canyon walls with shuddering effect, like a sonic boom. The bulldozer sank into the darkness of the cold subsurface waters, its dim shape of Caterpillar yellow obliterated, after a second, by the flare of an underwater explosion. A galaxy of bubbles rose to the surface and popped. Sand and stone trickled for another minute from the cliff. (127–8)

In an ensuing chase, Hayduke levers boulders into the road to impede the police, one of whom is driving a deluxe new Chevy Blazer. The boulder finds its mark:

> There was an anguished crunch of steel as the Blazer, squirting vital internal juices in all directions—oil, gas, grease, coolants, battery acids, brake fluids, windshield wash—sank and disappeared beneath the unspeakable impact, wheels spread-eagled, body crushed like a bug. The precious fluids seeped outward from the squashed remains, staining the roadway. The boulder remained in place, pinning down the carcass. At repose. (138)

The metaphor of the crushed beetle implies that the machines are vermin that must be exterminated before they bring pestilence, and that inorganic nature, in this case a boulder, does an elegant and thorough job as a terrible swift sword of natural justice.

On a slow night, Hayduke stops off at a favorite haunt, the construction site, to monkey around:

> Full throttle forward. The tractor lurched ahead one turn of the sprocket wheels and stopped. The engine block cracked; a jet of steam shot forth, whistling

urgently. The engine fought for life. Something exploded inside the manifold and a gush of blue flame belched from the exhaust stack, launching hot sparks at the stars. Seized-up tight within their chambers, the twelve pistons became one—wedded and welded—with cylinders and block, one immoveable entropic white-hot molecular mass. All is One. The screaming went on. Fifty-one tons of tractor, screaming in the night. (244)

And one more for good measure:

Twilight silhouettes, blurred by dust, Hayduke and Smith stand at the controls of their tractors, peering forward. Then very quickly they climb off. The tractors go on without human hands, clanking like tanks toward the canyon, and drop abruptly from sight. The tanker, the shed, the billboard follow. Pause for gravitational acceleration. A bright explosion flares beyond the rimrock, a second and a third. Bonnie hears the thunderous barrage of avalanching iron, uprooted trees, slabs of rock embraced in gravity, falling toward the canyon floor. Dust clouds rise above the edge, lit up in lurid hues of red and yellow by a crescendo of flames from somewhere below. (318)

Variations on a theme of industrial sabotage, these four scenes of ecoterror porn meld violence and suffering with spectacular, erotic action. Inanimate brutes die like animals, squirting, seeping, screaming, sparking, and seizing. Oil is their lifeblood, among other fluids. The pleasure with which Abbey details their demise registers their violent power to devastate a landscape to fit it to human needs: live by/die by. The gang has no compunction about despoiling the lake or the canyon with machine detritus. Like Thelma and Louise, who met a similar fate to the machines, everything over the edge is irrelevant. What matters is the battleground of the construction site, where the daily inroads against wilderness may be slowed by zealous counterpunches. These tactics of desperation contrast the failed legal alternatives to lobby, sue, and pacifically protest.

Abbey records an era in American history when the energy of petroleum had given the engineer unlimited power to alter even the most recalcitrant lands, which aligned perfectly with a national agenda of industrial development. Against the engineer's "we can" and the government's "we will" comes the environmentalist's "should we?" Daisies in gun barrels be damned; Abbey fights The Man on his own destructive terms. The Hollywood spectacle of dying metal giants merely gussies up the brutal guerrilla tactics of a futile tribe. New machines chuffing diesel fumes queue for work.

Yet Abbey's fringe ecoterrorist ethos continues to inspire conservation groups around the world. By refusing to operate within the rules of the master's house, rules involving permitting, regulation, and litigious means of counteraction, Abbey's gang of apes defined for a rebellious 1970s generation the kinds of violence that might be justified means against a violent regime of industrial development: don't kill other humans; kill their oil-fed enablers and wreck their habitat of roads, dams, and bridges. At a time when the majority of environmental action was poignantly nonviolent and even passive, Abbey's bellicose actions use the master's tools to dismantle the master's house.

John McPhee is an environmental writer with more sympathy for people of all political stripes. He consistently succeeds in identifying the humane at work within various environments. *Coming into the Country*, perhaps his best book, contains a vignette that provides great insight into the mindset of Alaskans, both native and transplant, in regard to petroleum. The extremes of Alaska's seasons and terrain highlight petroleum's power to broaden the survivable range:

> The forest Eskimos' relationship with whites has made them dependent on goods that need to be paid for: nylon netting, boat materials, rifles, ammunition, motors, gasoline. Hence, part of the year some Eskimo men leave the river to find jobs. These pilgrimages to the wage economy are not a repudiation of the subsistence way of life. They make money so they can come back home, where they prefer to be, and live the way they prefer to live—foraging the wild country with gasoline and bullets. (37)

There is no judgment here, regarding either whites or Eskimos, wages or subsistence. McPhee shows us a glimpse of the superhuman power lent by guns and gas, which help even well-adapted native cultures gain a firmer grip on existence at the margins. McPhee does not harp on the unnecessary goods a wage economy also provides, like cigarettes, alcohol, and nudie magazines. He shows the dignity of natives who accept alien work and money in exchange for practical and valuable advantages: better fishing, hunting, and transportation. They choose the point of engagement with white culture, and that point is gasoline. It makes us superhuman, which a human would like to be when facing winter at seventy degrees north latitude.

Another perspective on fuel in the wild country comes through Jon Krakauer's breakthrough essay "Death of an Innocent," the 1993 magazine story that preceded his book and Sean Penn's movie adaptation, *Into the Wild*. Gloss: an upper-middle-class boy from a Washington, DC, suburb, Christopher McCandless, graduates from college, drops off grid, and goes rogue in the American West, sojourning in the canyon lands and desert for two years before heading out on his final foray into the wild. Well, fairly wild. Fifteen miles from

State Route 3 near Healy, Alaska, across one notable river, he lived in a broken-down bus for about four months, on a trail blazed in mid-century by a mining company. Using the moniker Alexander Supertramp, McCandless rolled at least seven thousand miles on rails and tarmac before hiking only twelve miles into the bush. His whole journey from graduation day in 1990 until his death in 1992, a demon-hunting soul-haunting odyssey, was a journey enabled by petroleum. Though Supertramp is called a "leather tramp" by his friends in Penn's beautiful, idealizing movie, indicating a foot-borne journey, he was in reality more of a "rubber tramp" borne by his thumb into automobiles.

How poignant, then, that his austere will to live in nature is so intimately entwined in the end with a broken-down vehicle. Ever since the film that framed Supertramp's portrait in tramp history, the Stampede Trail has been frequented by pilgrims seeking the "Magic Bus" where Supertramp scrounged out survival and found scraps of his Shangri-La, but by mid-summer succumbed to hunger. He died lying supine, looking at the sky, in the back bunk of the bus. His primordial escape back to nature is actually a modern industrial story—down to the very existence of the dead bus that fixed Supertramp too long in one spot.

According to Krakauer, Supertramp's bus hosted diverse scenes in its history. It was first a Fairbanks city bus, then a transport wagon for mine workers towed by bulldozers, then a broken-axled shelter for backcountry travelers, then Supertramp's roost and coffin, and now a site for wilderness cult curiosity. The bus is a strange hulk deposited in a land of enduring wildness. Its incongruity in the landscape is the major ingredient in its magic. Supertramp's self-fashioned "final and greatest adventure" and "spiritual pilgrimage" has been explored existentially more than geographically, as interpreters have turned him into a darling of anti-materialism, devourer of wilderness texts, philosopher of postindustrialism, wanderer and dreamer, great outdoorsman (90). The bus attenuates each of these personae.

Supertramp himself recorded no irony regarding his reliance on the big metal junker, even though he had purposely left behind basic survival gear like a compass, a map, and a tent. The bus was magic; at least it was *for him*. It was part of his narrative of triumph and reward for tackling adversity. Soon after encamping under its roof, Supertramp carved his mini-manifesto (it would fit in a few Tweets) into a scrap of plywood, words that would become his eulogy. It ends: "No longer to be poisoned by civilization he flees, and walks alone upon the land to become lost in the wild" (90). His triumphal photo is a selfie in which the entire background is the metal side panel of the bus; not Denali, which he could see from his roost; not his library of Tolstoy and London and edible plant guides; not the riparian rocks and scrub pine where he toiled through summer days.

As reflected in both Krakauer's and Penn's accounts, Supertramp's journal reveals absolute glee at finding his magic bus in the bush, a welcome shell to cover his exposed body. He finds the bus only eight minutes into the film. Howling with wolfish abandon, he beats the mattress, flicks the lighter, burns the litter, and lights up the wood stove before tucking into bed to the light of a gas lamp. Immediately he gets comfy in his into-the-wild-indoors. None of us can blame him for the true animal satisfaction of finding a warm dry safe place to sleep, much as it tarnishes his prefigured resolve to purify himself of civilization. Lost in context, the bus no longer represents society, industry, history; now it signifies something much simpler and more beautiful in his eyes: staying alive. The magic bus montage equals the highest emotional moments in Penn's optimistic film, those inspired by mountains, caribou, romantic love, and the choreographed epiphany of his death.

Krakauer's and Penn's protagonist is an intelligent, playful, caring young man with a clear sense of purpose and an implacable ethic. His life and death are mutually fulfilling and even celebratory in the film, where he dies imagining the embrace of his reformed parents. The literal landscape belted by snowy mountains blends freely with the plywood narrative of the "aesthetic voyager" preparing for a "climactic battle to kill the false being within and victoriously conclude the spiritual pilgrimage" (90). Cultural critics, however, have found the Supertramp story to be less of a triumphant tragedy and more a terminal case of the muddling symptoms of consumer petroculture ennui. Timothy Morton sees his story as "fatal experimentation with masculine Nature" based on a fantasy of control and order created by the very society Supertramp sought to kill by abandonment (*Queer Ecology* 280). A lifetime of material privilege has two opposing effects: disdain and dependence. Over-access to stuff, in conjunction with an unhappy abusive family, infuses in him a visceral abhorrence for the so-called poison of civilization. His revulsion actuates the myth of the self-made man, accountable to nobody, independent from human ties like love, empathy, and vulnerability (280). But his dependence on industrial culture has only been extended by the oxymoronic wilderness bus, which substitutes for the adopted road-home he had previously found with wandering hippies in their rainbow camper. The magic bus is fixed in place, going nowhere while it figuratively drives Supertramp through his dreams and nightmares.

Supertramp lives in self-imposed ignorance of his actual surroundings. Without a map, his dwelling in the environs of the bus becomes a mythical place of his own creation. The magic bus is the center of a tight circle, and the center must hold for him to keep his bearings. Several weeks in the wild seem to have been enough to cure him of ennui. His terse journal records that his attempt to return to town in July 1992 was foiled by the swollen Teklanika River. Unburdened by maps, he was unaware of two easy recourses, one a river crossing cable only a quarter mile upstream from his attempted crossing, the other, a cabin stocked with provisions for backcountry rangers about six miles south of the bus. Ignorance of these aids made him over-reliant on the bus as shelter, drawing him back to that magic locale that, in his increasing desperation, he came to see as a trap. The far side of the river and beyond, the road and town, retreat from the margins of his mind map.

His declining health closes the circle further. His refusal to plan, his renunciation of the existing geography in preference for a personal mythology, is enabled by the magic bus. Rather than spending the summer wide-ranging across a remote area, as thousands of backcountry explorers do every year in American wilderness, Supertramp circled tightly around a derelict bus, squabbled with his inner demons, and miscalculated reality in terms of time and space. His moments at peace with pure nature respected, this story is a pyrrhic victory in which the real vastness of Alaska lies undiscovered beyond the rusting hull that is at once a defunct bus and a demon-haunted mind.

His death is explicable not just by exposure or starvation or poisoning, but in essence by his equation of freedom with lack of planning. It was an equation that had worked out many times before during his ranging in the lower forty-eight and Mexico, where he was frequently bailed out of trouble by generous people in cars and motorboats. The magic bus temporarily assuages his failure to provide for his time in the wild. Then, inadequate in itself, his addiction to the bus kills him. Central Alaska was wild enough that he had not seen another human all summer, and help would not arrive until a few weeks after Supertramp's sojourn had come to an end.

On September 6, 1992, an astonishing six people in three separate parties coincided at the magic bus, some of them driving the gassed-up all-terrain vehicles that have reconfigured our definition of wilderness. There they found the remains of McCandless. Now a memorial site, the magic bus still squats in that spot, lower and rustier by the year. It remains an enduring symbol of the incursion of industry into wilderness. Literally getting out of gas-powered machines to step into wild country is easy. But getting outside of the petroculture mind-set was more than McCandless could accomplish in his short life.

IV Oil's locales

Oil created some of the richest people the planet has ever known: J. D. Rockefeller's
adjusted wealth was about a half trillion dollars. But the trickle down to the
subordinate workers in the oil economy was slow indeed. Sitting behind a massive
desk in New York City, Rockefeller had little exposure to the matter of oil itself: the
accoutrements of drilling, refining, and filling, and the economic chanciness of
selling a commodity only a few cents above wholesale price. Instead, poets have
gone to these places. In a poem titled "Filling Station," Elizabeth Bishop notes
the indignities of roadside mechanical work versus the perennial human desire
for self-improvement, often displayed through possessions. The poem begins,
seemingly, as a snobby anthropology of the gas station men:

> Oh, but it is dirty!
> —this little filling station,
> oil-soaked, oil-permeated
> to a disturbing, over-all
> black translucency.
> Be careful with that match!
>
> Father wears a dirty,
> oil-soaked monkey suit
> that cuts him under the arms,
> and several quick and saucy
> and greasy sons assist him
> (it's a family filling station),
> all quite thoroughly dirty.

(ll. 1–13)

As matter, oil is powerful outside of the fuel tank as well as inside due to its
stickiness, viscosity, indelibility. It permeates its sphere with a "disturbing, over-
all / black translucency"—disturbing because it is both black and bright, and it
stains. This little stop-n-go world is completely defined by the aversive, alluring
duality of gas grease. The actual business of oil dehumanizes its workers into
monkeys, and the family business that is the icon of proud middle America
is here a trap where sons are chained in prosaic servitude, stuffed in greasy
overalls alongside their ragamuffin father. In response, the sons are "quick and
saucy," traits that might be admired in ambitious young men, but Bishop cancels
the virtue by repeating "greasy" as their leading trait. Whatever they may be
underneath, it is hard to perceive beneath all that filth.

Still, Bishop presses past the surface, as a poet must, to come away with some insight into the people she stereotypes as grease monkeys. Presumably, as they attend to her car, she noses around their property:

> Do they live in the station?
> It has a cement porch
> behind the pumps, and on it
> a set of crushed and grease-
> impregnated wickerwork;
> on the wicker sofa
> a dirty dog, quite comfy.
>
> Some comic books provide
> the only note of color—
> of certain color. They lie
> upon a big dim doily
> draping a taboret
> (part of the set), beside
> a big hirsute begonia.
>
> (ll. 14–27)

The wicker, the dog, and the doily are shades of a better life, but they have succumbed to the universal grease. This is a poem about color and color's diseases— how the world can be suffocated by the slick-sticky dirt of oil—and the things that escape this fate, gesturing to another life in this oil-scape: comic books (low-class entertainment, but colorful!) and, most remarkable of all, a living plant that would have exuberant pink blooms to shout out against the chromatic din. For Bishop, these exceptions prove a higher value and order to a thoroughly dirty world. Hope is not lost; in fact, through the doily and the plant, love shines through:

> Somebody embroidered the doily.
> Somebody waters the plant,
> or oils it, maybe. Somebody
> arranges the rows of cans
> so that they softly say:
> esso—so—so—so
> to high-strung automobiles.
> Somebody loves us all.
>
> (ll. 34–41)

This is a gently feminist turn, where domestic accomplishments like embroidery and horticulture spruce up the grease-scape and show by their very existence that care and pride ground this goopy place. The dog is cozy. The plant survives and blooms. The absent presence is the maternal caregiver, whose several sons and husband sling grease and gas to fill themselves with milk and meat.

Most importantly, this filling-station world exists to cater to the needs of our new favorite pets, our cars, and the line of perfectly arranged Esso cans whispers a lullaby to those machine-babies suckling oil. From an initial aversion to obvious filth, the poem comes around to rejoice in a universal succor enjoyed by animal, plant, and machine. There is dignity in this work. Like the charcoal man smeared in his soot garb, the filling station men help us. Unlike the silk-suited oil tycoon, they wear the marks of their labor.

Elizabeth Bishop wrote one other oil poem worth mentioning here. "The Fish" is mostly a visual exploration of the body of an old, battered lake fish hauled aboard a rented rowboat. But the culmination is notably petro-philic:

> I stared and stared
> and victory filled up
> the little rented boat,
> from the pool of bilge
> where oil had spread a rainbow
> around the rusted engine
> to the bailer rusted orange,
> the sun-cracked thwarts,
> the oarlocks on their strings,
> the gunnels-until everything
> was rainbow, rainbow, rainbow!
> And I let the fish go.

(ll. 65–76)

Unlike the dark, greasy filling station where light is swallowed, here oil emulsifies on water and reflects all colors. It turns pollution and decay into a fragile beauty, both alluring and repulsive, something Bishop's images often pursue. The rainbow vision propels her mood from the triumph of victory (she caught the venerable old fish), to the empathy of mercy. Like "The Filling Station," an oily aesthetic decorates the scene with an equivocal chemical nature that fans out in a range of color-plays from black vacuum to shimmering rainbow. Oil's tawdry company—grease, bilge, rust—do not ruin its magic materiality.

More contemporary poems look back on the legacy of the cheap oil era of the twentieth century. Philip Appleman's 1996 poem "How Evolution Came to Indiana" snipes at a regressive statewide biology education policy while also meditating on the fate of the obsolete dinosaurs of the mid-century car industry. Roaring through "the swamps of Auburn" and "the jungles of South Bend," these hulks are now metallic fossils that show in their forms a cause for extinction: "3-speed gears ... 2 chain drive" (ll. 4–6)—inefficient engineering. Appleman quotes Darwin:

> *There is grandeur in this view of life,*
> *as endless forms*
> *most beautiful and wonderful*
> *are being evolved.*
> And then
> the drying up, the panic,
> the monsters dying:
>
> (ll. 16–22. Emphasis in original)

In the old world awash in fossil fuels, these paleolithic cars ruled the landscape, triumphed in their excess of raw power and battled to be alpha-dinosaur in gasoline's golden age. Most of these brands were extant in the first half of the twentieth century and manufactured in Indiana, supporting the small economies of company towns for a human generation or two. Like Ankylosaurs, Hadrosaurs, and Prosauropods, these rusty relics are hardly recognizable to youth, who look to more "successful" forms of cars—including some makes that coexisted with the Deuseys and Stutzes, like Cadillac and Ford, as well as new competitors with better mileage and reliability like Honda and Toyota—that filled the niches left open by extinction. As any economy is inherently an ecology, we can say that these cars thrived in the "wet" conditions of the American oil boom, then declined through "the drying up, the panic," caused by a change in environment.

Appleman lists the mechanical traits that he characterizes as failed experiments in engineering. As in evolution, though, extinction often cannot totally be attributed to a flaw or failure in the dying species; biological and economic evolution are both racked with happenstance and bad luck. Even if biological evolution was not taught in Indiana classrooms, its citizens need only look in their junk yards and tall-grass ditches to see the principle hold true. In evolution, there is often progress (at least when perceived in retrospect), but there is always ruination.

Appleman quotes Darwin's final passage on the "grandeur in this view of life," but rather than riff on Victorian triumphalism, his poem is an elegy about the end of innocence during the early era of black gold. Younger generations may feel a similar twinge for the more recent hulks of the American car industry, endangered by their ruinous appetites during the 1970s oil embargo—the Cadillacs and Lincolns that became "hoopties" of hip-hop fame. Nostalgia is a powerful potion, and each generation suffers through its own battle between self-indulgence and frugality. The Hummer is the leading contemporary example of an outlandish vehicle made extinct by more efficient competition. The Hummer's open flouting of efficiency, in fact, made it a laughing stock and soon a discontinued brand. It took fuel bravado to an extreme, and American society discovered the limits of it tolerance. Cars fight these evolutionary economic struggles on the ecological plain of gasoline.

For the poet Sina Queyras, the interstate offers a mosaic of disjointed images that correspond to our cognitive experience of landscape experienced via automobile. Intense with enjambment, the images in "Endless Inter-States" (2009) jumble together without consistent form, threatening to wreck, but the poem veers between the spacing of regular tercets. Before surging down the road, Queyras pauses to remember past travelers and their metal carapaces. Like Appleman's poem, the past clutters the roadside in a mechanical version of natural history: cars are like bugs of various species, all identified and studied for their curious features. Simultaneously, in a fast-paced consumer economy, they are ugly garbage, blithely discarded and ignored.

Sagebrush, dead-ending on chain
Link, old cars collecting like bugs
On the roadside, overturned, curled, astute,

Memory of the Overlanders,
Optimism, headlong into
Hell's Gate. Churn of now,

The sound barriers, the steering
Wheel, the gas pedal, the gearshift,
The dice dangling, fuzzy,

Teal, dual ashtrays, AM radio
Tuned to CBC, no draft, six cylinders,
The gas tank, the gearshift, easing

Into the sweet spot behind
The semi, flying through Roger's
Pass; the snowplow, the Park

Pass, sun on mud flap, the rest stop
Rock slides, glint of snow, the runaway

Lanes, the grades steep as skyscrapers,

The road cutting through cities,
Slicing towns, dividing parks,
The road over lakes, under rivers,

The road right through a redwood,
Driving on top of cities, all eyes
On the DVD screen,

All minds on the cellphone,
The safari not around, but inside
Us: that which fuels.

(ll. 117–37)

From a view of the wasted roadside, Queryas passes into the domestic amenities of the car's interior. Our cars are rolling living rooms appointed to entice us to stay behind the wheel, to feel the powerful engine as an extension of our own bodies. From that position of power, everything outside, however naturally magnificent, is diminished by its access and exposure. The road cuts, slices, and divides, rather than uniting places, as we often assume. The road scars landmarks in a comprehensive array: towns and cities; parks, lakes, and rivers; forests filled with massive redwood trees. The road even cuts through individual redwood trees.

The poem ends by erasing even those sketches of the world. The outdated experience of sightseeing by car, a kind of "safari," observing the terrain's accessibility and diminishment, marveling at the engineering and money that supports such infrastructure, has now melted into a technological solipsism. Existence in the early twenty-first century has turned our eyes to screens and minds to virtual spaces. A brain moves through actual space, but it thinks of immaterial abstractions. That which lurks beyond the screens, outside the windows, has become irrelevant, as abstract and aphysical as a "like" on Facebook. And less interesting to many.

Gentle folk in the eighteenth century who traveled by carriage through the Alps were known to draw their curtains against the mountains' ugliness, seen as evidence of God's displeasure with mankind. From the world's greatest vistas, their world then retracted into a space between carriage windows, about four feet square. According to Queryas, today we travel with the curtains of our minds drawn against the varying landscapes, for even glimpses of vast mountain ranges at high speed are less enthralling than these self-oriented world-simulations within our new favorite prosthetic, the smart phone. That phone, plugged into the car's power stream, runs on gas, too.

V Pastoral oilfields: The suburbs

Suburbs are locales uniquely beholden to fossil fuels. Their existence depends upon a daily commute into and out of them, originally by train, but now mostly by one-person cars. They are literally and aesthetically located between rural and urban spheres, promising the leisure of a pastoral idyll and the plump salary of an urban white-collar professional. Early American landscape designers like Frederick Law Olmstead were inspired by English urban parks to unite the culture and wealth of the American city with the supposed tranquility and purity of the rural sphere (Gill 21–2). Contemporary accounts of the design principles of early suburbs allude to the satisfaction felt by residents when raising chickens and growing their own vegetables, and usually prescribed a mixed-use flower and vegetable cottage garden cultivated with an ethos of self-sufficiency and honest labor.

These Edenic edicts align with the early positive perceptions of the first American suburbs, glorified as a proving ground for pioneers on a new frontier of naturalized domesticity. Along with independence and self-reliance, the early suburban model tested the viability of starkly gendered ideas of space and time that would survive for more than a century (Gill 22). In 1869, Harriet Beecher Stowe praised the railroad as a vector for achieving the best of both worlds because they enable "men toiling in cities to rear families in the country" where heads of family may "seek a soil and climate which will afford ... outdoor labor for all" (qtd. in Gill 23). These railway suburbs soon gave way to the twentieth-century version based on sublime automobility—an individual's escape from muscle space into fossil space, which dedicates the body, the mind, the home locale, and the family economy to a civic architecture designed around the car (Johnson 54).

Suburbs carry stereotypes that have been in play ever since the rapid construction of suburbs along rail lines in Europe and America in the 1800s. Rapid mass production of neighborhoods is thought to breed conformity and normativity. Space is highly privatized but often adorned in the appearances of welcome: picket fences studded with flowers, welcome mats next to home security signs, neighborhood groups mostly dedicated to vigilance against outsiders. The "gated community" is an updated, higher level of exclusivity operating on the same parameters as the earliest suburbs of the nineteenth century. Suburbs are severely landscaped, turning their rural or wild substrate into chartered rectilinear plots where a small variety of plant species foisted by the local garden center appear in a recursive infinity down the street. New subdivisions often adopt the name of the landscape they destroy: Deer Chase, Whispering Pines, Sunny Meadows.

Promised as the country house of everyman, suburbs are often bourgeois, culturally homogeneous, beholden to mass media entertainment (rather than live and local productions), and tend to house uniform racial and socioeconomic groups. They intensify consumer desire by providing the space to accumulate things such as televisions, cars, riding mowers, trampolines, patio sets, and display a voyeuristic layout to make constant neighborly comparisons.

Wendell Berry, philosopher of the virtues of muscle energy, laments how the modern home, especially the suburban one, "is a veritable factory of waste and destruction. It is the mainstay of the economy of money. But within the economies of energy and nature, it is a catastrophe. It takes in the world's goods and converts them into garbage, sewage, and noxious fumes—for none of which we have found a use" (*Unsettling* 52). Like the fossil-fueled modern farm of monoculture and high-concentration animal husbandry, the suburban home exists outside biodynamic material loops that turn waste into fertility. This is both a material and ideological point: Berry argues that when people lived where they worked—in cottage industry, for example—dwelling and productivity were mutualistic. Place was indistinguishable from purpose, and each dwelling was uniquely accommodated to its dweller. But in the era of the "industrial conquistador," Berry writes, "everything around him, everything on TV, tells him of his success: *his* comfort is the redemption of the world," but that emblem of success is abstracted from the consciousness of work performed to succeed (53). Suburbs drive a wedge between work and life, selling us the notion that we *do* only so that we may *have*. To simply *be* does not compute in this economy.

Even in its glory days of the 1950s, the gender-determined suburb became an intriguing place of dissipation: bored housewives spewed venomous gossip; sheltered children, also bored, sought outlets in virtual violence, drugs, and sex; absent working fathers kept *pied-à-terre* apartments in the city and lingered there suspiciously. Equipped with a car, the wife was able to redefine herself from being a producer of food and clothing in a traditional domestic sphere, to becoming a consumer of carefully marketed brand goods: canned food, clothing, cosmetics, cars, washers, and anything else sold in the petro-marketplace (Nye

181). The suburb is the classic industrial-era locale for mistakenly equating personal value with possessions, rather than actions, accomplishments, or experience. In stereotype, they are inane utopias occupied by self-conscious but unremarkable people.

After the apex of idyllic suburbery in the 1950s, the time in America when consumerism was a normative patriotic lifestyle, the paint started to peel off the country-kitsch picket fences. When the inner suburbs around major cities grew shabby or crowded or too expensive, developers conscripted the next ring of farmland to residential service. Commutes grew longer; roads, larger; cars became more like living rooms than rolling chairs. The evolution of the suburb in the twentieth century entailed a deeper investment in its worst characteristics— inefficient land use; loss of personal time and community cohesion; perhaps most apparently, flagrant waste of fuel in cars and lawn machines.

Why would anyone live there? Because suburbs promise amenities that many people find attractive, offering comfort, though along with smidges of shame and inconvenience. Many humans like to live with people who look and talk like themselves, have similar backgrounds and common dreams. Suburbs offer an illusion of control that results in painful contradictions. They provide safe space for children to play (if only we could get a stop light at the corner of Holly and Mulberry). They are leafy, flowery, wormy (and we really must spray the dandelions this year). They are quiet (Mr. Jones is mowing again!) and safe (Maggie saw that same ugly man loitering on the corner last night). Houses are capacious and stately (I almost keeled over while cleaning the great room today).

These suburban contradictions are intimately related to their paradoxical nature: they are pastoral oil fields. Designed to be rural, they rely on cheap abundant energy to be viable. Seemingly existing as both at once, they are truly neither. They are faux pastoral, because they involve organic nature but fail to produce rural goods like food and raw materials. They are faux oil fields because they are rich in oil, and defined by a dedication to oil culture, but they consume oil rather than produce it.

In all their liminal, in-between identity crises, suburbs are actually quite interesting to writers. John Updike, noting the marginal spaces where he often sets his novels, explained his interest in "middleness with all its grits, bumps, and anonymities, in its fullness of satisfaction and mystery." In a 1966 interview he said, "I like middles. It is in middles that extremes clash, where ambiguity restlessly rules" (qtd. in Gill 155). Television shows like *Picket Fences, Six Feet Under, Desperate Housewives*, and *Weeds* have leitmotifs of suburban inanity. The cultivated outward appearances belie the illegal, rancorous, and raunchy doings of the characters within. *Weeds*'s theme song is Malvina Reynolds's 1962 satire of the suburbs, "Little Boxes," which lampoons the "ticky-tacky houses" that all look alike. While the song plays during the opening credit sequence montage, we see a train of identical SUVs drive down the street.

Film culture has drunk the potent cocktails of the suburban gothic, the conflation of utopia with dystopia: the *Nightmare on Elm Street* slasher series; Marty McFly's oxymoronic Hill Valley, seen both in the past and in the present, in *Back to the Future*; the deaths of innocents in houses shaded by trees devastated by Dutch Elm disease in *The Virgin Suicides*; the sarcasm, self-indulgence, and self-hatred displayed in the suburban culture of *American Beauty*.

Literature of the suburbs predates the lavish vapidity of the suburbs depicted in television and film. The suburban gothic of America's late twentieth century had its predecessors two centuries earlier, as the world's great cities, London and Paris, began to develop outlying rural terrain to accommodate industrial-era populations and the coal-driven train system that transported workers from country to city and back again. In the early decades of the 1800s, London sprawled outward in often ill-conceived rapid development. In the nowhere/now-here bonanza, the space between old London and Hampstead Heath was filled with Camden Town, and new constructions in Stockwell, Brixton, Maida Vale, Kilburn, and Holloway soon followed (Witchard 24–5). Novelist Molloy Westmacott likens this development with dark magic in *The English Spy* (1826): "The rage for building fills every pleasant outlet with bricks, mortar, rubbish and eternal scaffold-poles, which, whether you walk east, west, north, or south, seem to be running after you ... It is certainly astonishing: one would think the builders used magic or steam at least" (72). If the builders were not wizards using magic, they were engineers using the alchemy of steam-driven engines. Both the demand for new housing and its rapid construction were driven by fossil-fueled industry.

Charles Dickens sees nothing but squalor in the environs of new train stations, which he imagines as a giant trampling the land: the place "is neither of the town nor country. The former, like a giant in his travelling boots, has made a stride and passed it, and has set his brick-and-mortar heel a long way in advance; but the intermediate space between the giant's feet, as yet, is only blighted country ... a pleasant meadow ... was now a very waste, with a disorderly crop of the beginnings of mean houses, rising out of the rubbish, as if they had been unskillfully sown there" (*Dombey and Son* 517, 519). Dickens's clever arrangement implies that the land is still too rural to consider the new things as buildings. They must be ugly crops. The country is ruined, and the city is impossibly remote. Where are we? In a new fuel landscape, built by coal, accessed by coal, and created out of the paradoxical desire to escape the coal-industrial habitat.

The disorder and squalor that characterized the new construction sites observed by Westmacott and Dickens usually domesticates, in time, toward the distinctive neo-nature of the suburb. By the early twentieth century, these suburbs near railway stations would be established neighborhoods nestled into their natural surroundings with elevated property values. The second wave of suburbs was enabled by the automobile. In 1910, two hundred thousand automobiles were registered in America; in 1919, six million; in 1929, twenty-three million. The traffic jam, then a new development, scrambled nerves between the world wars. This stunning growth in car ownership, made possible by cheap domestic petroleum, caused the infill of car-only developments between the older suburbs already established along rail, streetcar, and trolley routes (Gill 26). Complete rings of suburbs around major cities such as London, Paris, New York, Washington, and Los Angeles, were in place before the start of the Second World War. We generally think of suburbs as a postwar phenomenon, but the railway and the car, along with the coal and petroleum that powered them, had already fundamentally changed the mode and locale of dwelling.

In the 1921 poem "The Tulip Bed," William Carlos Williams examines the complex boundaries between a built environment and nature, the nexus between pastoral and urban spheres that the suburbs represent. After depicting the May sun's power to unfurl the leaves, Williams observes:

> Under the leafy trees
> where the suburban streets
> lay crossed,
> with houses on each corner,
> tangled shadows had begun
> to join
> the roadway and the lawns.
> With excellent precision
> the tulip bed
> inside the iron fence
> upreared its gaudy
> yellow, white and red,
> rimmed round with grass,
> reposedly.

(ll. 8–21)

Williams vexes the grid of neighborhood streets with a tangle of branches and shadows. The insistent geometry of the roadway yields to an organic mesh of boughs and grass lawns. But the foliage is precise, too; the tulips are laid to present a premeditated appearance of lush order, contained by the delimiting fence and classic "perfect" one-species grass lawns. Though this is closer to an image poem than a meditation on the nature of suburbia, Williams captures the energies of the wild, especially shadows and tulips, butting up against the designated boundaries of a closed space. Chaos and control vie for dominance. The literal begs for metaphor: shadows might be housewives; tulips, suburban children. Everything meant to be owned and contained, presented just so, is in revolt, and the gaudy tulips nip at the iron fence as the rain inevitably rots it away. The poem, provisionally an idyll of suburban contentment, squats over a dark well of equivocal energies. Petroleum is the energy that organizes the space into "crossed" streets and designates the placement of corner houses and the rest of the array: grass, fence, tulips. But the vying entropic energy of organic wildness will never be squelched out. This little vignette of a suburban poem contains a war of the worlds: industrial fossil-fueled civilization versus sun-and-soil-fueled nature. The last word, "reposedly," suggests the appearance of things rather than their reality—the ultimate suburban condition.

Phyllis McGinley's 1941 sonnet "The 5:32" catches the essence of the suburban woman's experience during the war years. Thinking about the legacy of her life, she identifies the most memorable time of day in the endless progression of humdrum days lived out by the "housewife-mother"—the woman triply domesticated and controlled. The housewife-mother's fixed daily moment is meeting the evening train and the city-working husband:

She said, If tomorrow my world were torn in two,
Blacked out, dissolved, I think I would remember
(As if transfixed in unsurrendering amber)
This hour best of all the hours I knew:
When cars came backing into the shabby station,
Children scuffing the seats, and the women driving
With ribbons around their hair, and the trains arriving,
And the men getting off with tired but practiced motion.

Yes, I would remember my life like this, she said:
Autumn, the platform red with Virginia creeper,
And a man coming toward me, smiling, the evening paper
Under his arm, and his hat pushed back on his head;

And wood smoke lying like haze on the quiet town,
And dinner waiting, and the sun not yet gone down.

(ll. 1–14)

This still-life poem is idyllic and creepy at once, reflective of a popular culture that has fashioned the suburbs as a setting for disturbing rituals and secret lives, obscured by the banal spectacle of hair ribbons, Virginia creeper, and conformity. At 5:32 p.m., the train releases the workers into waiting cars, fossil fuels shuffling them along various legs of their quotidian journey. These lives, it seems, are not their own but instead are directed by the rhythms of transportation: children and mothers confined to waiting cars, men wearily closing out another day of hither and thither. Dinner waits on every stove up and down the street. The sun illuminates rural wood smoke. The only fissures to be found in this settled contentment are the bored scuffing children and the mother who, in spite of herself, cannot help but imagine her "world torn in two, / Blacked out, dissolved." Though the poem is about the warm comforts of a suburban routine, apocalypse is the opening catalyst that shows the frailty of it all. It may also serve as the woman's subconscious fantasy of something, anything else.

If McGinley's poem is deliciously ambiguous, insinuating the apocalypse in the idyll and punctuating stability and rhythm with violent change, Louise Bogan's "Evening in the Sanitarium," also from 1941, dispenses with the niceties of suburban appearances in order to study its psychoses. The poem's chilling echo of "Jim home on the 5:35" indicates women destabilized by isolation and conscripted into domestic service:

O fortunate bride, who never again will become elated after childbirth!
O lucky older wife, who has been cured of feeling unwanted!
To the suburban railway station you will return, return,
To meet forever Jim home on the 5:35.
You will be again as normal and selfish and heartless as anybody else.

There is life left: the piano says it with its octave smile.
The soft carpets pad the thump and splinter of the suicide to be.
Everything will be splendid: the grandmother will not drink habitually.
The fruit salad will bloom on the plate like a bouquet
And the garden produce the blue-ribbon aquilegia.

(ll. 14–23)

Though each woman—bride, older wife, and grandmother—has a different story behind their sojourn in the sanitarium, their emotional topographies have become uniformly flat: no elation, no bitterness, no immoderation. In order to pull off such feats of indifference, Bogan releases her women into a general society as "normal and selfish and heartless as anybody else." They are considered crazy, marked by their nonconformity, but the system promises to return them to normalcy—to the land of perfect salads and exceptional gardens.

Jim, who escapes on the train every day, is nothing in this poem but another chore. Both poems were published in 1941, a year notable as the beginning of a revolution in the American way of life, with war radically rearranging settled patterns of domesticity. McGinley's apocalypse may be America's commitment to the war. The men on the evening trains might any day go abroad as soldiers; the women waiting in cars may become versions of Rosie the Riveter. The suburbs were floating downstream through a four-year rapid, after which, in the 1950s, their hegemony as models of development would be reestablished.

As a proxy of rural dwelling, the suburban ideology mystifies its reliance on machines and fuel. But the real agricultural sphere that produces crops and meat has also changed its entire structure to benefit from the energy of fossil fuels. No longer biodynamic closed systems in which waste from animals became fertilizer for crops, petroleum has thrust itself into the cycle and shut it down. In place of animal labor come machines; in place of natural fertility, artificial petro-based fertilizer that reduces complex plant nutrition to three elements sold in bags: nitrogen, phosphorous, and potassium. Not only do the suburbs pose as something they are not, productively rural, they pose as something that hardly exists anymore: the cozy small farm that is at once its own ecology and economy.

Small wonder that futurists envision the suburbs of the twenty-first century in two basic forms: unevolved, they may become isolated ghettos with no economy or services and rotting infrastructure; or they may evolve into post-suburban communities retrofitted with town centers and local businesses to reduce dependence on petroleum. Suburbs will either stay dedicated to a dissolving ontology based on an industrial idyll, or evolve into mini-cities that can be navigated using our own muscles. Suburbs of the future might encircle their economy by dedicating the surrounding land to small-scale agriculture: every yard a productive garden, every tree-lined street an alley of fruit-yielding usufruct. To survive, suburbs will have to become both more rural and more urban at once. They will have to repudiate the original vision of neither/nor and embrace a both/and flexibility. Many American suburbs, including the one I write in, built in the 1950s, are now actively interpolating blank spaces of grass and parking lots with higher-density mixed-use zoning, public transportation, and community-based agriculture. The suburbs of the future will have to make do with a lot less oil. Without petroleum, we must stay closer to place most of the time. These energetic limits may generate new value and production in the middle spaces of suburbs.

VI A note on natural gas

We have explored two chemical states of fossil fuels, solid and liquid, and one remains. The dark horse fossil fuel, natural gas, has been historically less important than its geochemical cousins of coal and petroleum. Widespread extraction and infrastructure to support natural gas supplies did not come about until the middle of the twentieth century in America, and a bit later in England. Its brief history belies its importance to our present and future economy. Natural gas widely fuels furnaces, electricity generators, and specialized locomotion such as new city buses. Recently it has received a lot of attention because of its comparatively low greenhouse gas emissions, its cheapness due to the extraction technique of hydro-fracking, and its abundance. Environmentalists see it as a fuel that may transition us from the dirtier fossil fuels to the cleaner, but more expensive and dispersed, renewable fuels derived from sun and wind. Some futurists see natural gas as the platform fuel that we must use, not to maintain the status quo but to build the infrastructure of a new economy based on renewables: our turbines, solar panels, and electric vehicles require intensive energy to produce. Natural gas is Janus-faced: it is climatologically a better option than petroleum and coal, but the booming use of hydro-fracking to attain it and the economic necessity to use this violent technique to achieve reasonable returns makes the modern natural gas industry anathema to many environmentalists.

But there is very little contemporary literature of natural gas. I will discuss one contemporary piece that engages with this fuel on the environmental level: Barbara Hurd's 2013 prose poem, "Fracking: A Fable." The poem plunges through deep geological time to gain perspective on the life and death of natural gas; the locale of interest is the Appalachian basin. Through the superhuman eons when "everything took forever," land and ocean was in flux, mountains jutted out and slowly eroded, and, most importantly, legions of plankton lived and died, settling into a soft rock-submerged mass that would be cooked for millions of years. Now comes formation:

> Meanwhile, down in the black ooze, remnants of those tiny creatures that had been held in the mud were shoved more tightly together, packed side by side

with sludged-in sediment, cemented together, cooked by the heat deep in the earth, and converted into hydrocarbons. Layer after layer of crammed-together particles and silt began to sink under the accumulating weight of the mountains that grew above. Wrung of its moisture, its pliability, its flow, the mud slowly, slowly, over millions of years, turned into gas-rich rock.

After millions of years of uncertain, directionless change, humans evolve into the scene in a violent punctuation. Our tools are purposeful and keen, our actions swift. Eternity encounters immediacy, and the sleeper wakes with a jolt. The fracking method comes onto the scene in a geological instant: drills plunge deep into the rock and maneuver for the best purchase on pools of compressed gas, forcing fluid cocktails into fissures to break up the ancient seabed. The quiescence of a blind process is jolted into productivity, the wells give up their wealth, and the world is "lickety-split forever changed." Humans, "whose footprints, so conspicuous and large, often obliterate cautionary tales," cannot predict the outcome of their actions; economic demand and technological competence are the only rationales. Hurd ends her excursion with a note on the nature of time: "And now, in no time at all, not everything takes forever any longer." Our thirst for fossil fuels alters the nature of time from the "forever" of an agentless world to the insistent "now" of the Anthropocene. As with Ruskin equating rapid train travel with killing space and time, here Hurd offers an epitaph on the death of the rock, and the annihilation of the eons that must pass for biomass to slow cook into rich fossil energy.

We are a lucky species, evolving from modest nomads into world rulers during a ten thousand–year string of stable weather that fostered the agricultural revolution; finding fuel resources at nearly every corner of the planet to build and subsist our cultures; innovating tools much more powerful than our bodies to renovate the earth to our specific needs. We are an even luckier cluster of generations, alive during this exceptional time of energy expenditure from fossil fuels unknown to earlier history and highly temporary in duration. In our oil ontology we live outlandishly, beyond ecological limits and past margins of energy previously imposed on earthlings.

The literature of this high-energy world is analogously manic: filled with exuberance, angst, and rage, self-absorbed in the mundane mastery that fossil fuels exercise over the material world. Coal and oil have changed everything in just a few centuries. World population has grown about seven times in the last two hundred years: from one billion to seven billion, with mid-century projections greater than nine billion. This exponential growth—the great departure from millennia of relative population stability—can be explained largely through the energetic contributions fossil fuels have made to food systems and economies of production. The food calories from large industrial farms are generated with swarms of hydrocarbon calories that constitute fertilizer and pesticides, power large machines, process and distribute food, and keep it refrigerated. In parallel, hydrocarbon calories make other forms of industrial production possible, with legions of worker/consumers toiling around a glowing core of fossil energy. We cannot separate our present selves, physically or ideologically, from the great orbital gravity that hydrocarbons exert at the center of our lives.

But have we donned the wings of Icarus? The outrageous trajectory of energy use since the Industrial Revolution has us flying too high, and, crash landing being an ugly option, we must innovate in-flight. Fossil fuels are unsustainable in both supply and disposal as the carbon sinks fill up and supplies draw down. The current interest in ultra-low-grade tar sands oil and the furor over its intensive and destructive extraction method shows that in the early twenty-first century we are already in the fifth act of the hydrocarbon drama. Oil companies like BP and Exxon produce marketing campaigns that re-brand them as renewable energy corporations with "green" credentials. But considering ample supplies of coal and natural gas, the energy crisis becomes as much an issue of morality as of supply and technology. As individuals and societies, can we choose to descend with some grace, by questioning the logic of unrestrained consumption? Or may crises of ecology and economy force a freefall?

The next, and final, section explores the literature of renewable energy. New renewables inhere to the moral imperative of downscaling consumption because they are dispersed and hard to capture compared with fossil fuels. While evolving technology promises easier and more energy, renewables today may well demonstrate the historical era in which the moral issue of consumption comes into step with the material issue of supply. The fossil-fuel flash must give way to the shine, the flow, the breeze: forces comparatively pacific, but regenerative and time-honored. Renewable energy in literature is as much nostalgia as it is prolepsis.

7

Interlude: Human Food

Our connection with energy is never more intimate than with food. We consume other organisms to fuel our biomass-burning bodies. Because food has so many rich and complex connotations involving tradition, ritual, hospitality, and artistry, and because it is so redolent to our many food-attuned senses, it is easy to forget that, stripped of connotation, food is simply fuel. To reduce it to this basic entity of energy is not to destroy its higher-level metaphorical life. Reducing food to fuel is the most direct way to reckon with the idea that our choices in food represent particular attitudes toward nature. When Wendell Berry declares that "eating is an agricultural act," he reminds us of a direct line between the heaping plate and the soil, now obscured by the massive, fossil-fueled industry of modern capitalist food production (Young 11).

Whether we're conscious of the agriculture of eating or not, our choices in food translate directly to which segments of the food economy we support: local or global; organic or conventional; whole or processed. The Green Revolution of agriculture after the Second World War utterly changed agrarian practices across the globe, effectively substituting muscle-driven labor and organic fertilizers for petroleum-driven machine work and synthetic fertilizers and pesticides. While this paradigm shift is credited with saving developing nations like India from famine, its environmental legacy is checkered, at best. The costs of high-yield conventional food production include loss of crop biodiversity inherent to traditional polyculture systems, and exponentially higher levels of energy consumption from fossil fuels, causing greenhouse gas emissions, dependence on chemical fertilizers, and exposure to pesticides. Neo-Malthusians see the Green Revolution as the fossil-fuel-driven force that has permitted exponential population increases beyond the carrying capacity of a globe that subsists on renewable energy.

By eating, we extend the chain of chemical conversions that begins with sunlight and proceeds through photosynthesis to plants—their edible leaves, roots, and fruits—delectable biomass. When we eat animals, the chain is a link longer, since animals convert rough biomass, often downright inedible to humans, into meat, milk, and eggs. Polyface Farm's Joel Salatin, the "grass farmer" made famous by Michael Pollan in *The Omnivore's Dilemma*, calls his work "mob-stocking herbivorous solar conversion lignified carbon sequestration fertilization" (Anderson and Slovic 128). All food is synthesized sunlight, and the biodynamic loop among the sun, inorganic elements of the soil, and microbes, plants, and animals is self-sustaining.

As the chain from sun to plate gets longer, the ecological burden of the food grows: only about 10 percent of biomass calories fed to an animal become meat; the rest is dedicated to metabolism. This 10 percent rule is a moot point when animals graze and browse on things we cannot digest, like grass and scrub. But the industrial meat system has long used agricultural commodities edible to humans, like corn, soybeans, and wheat, to fatten animals for the slaughter. In this system, a hundred calories of corn yield ten calories of sirloin, so the meat represents a luxuriant use of biomass calories. Eating steak is like burning diesel: it is high-energy and high-input fuel, and the act is unsustainable in the sense that industrial meat factories consume many times the calories they produce. A large portion of those calories come from conventional feed corn energized by fossil fuels as well as photosynthesis. Eating blackberries foraged from the woods is like plugging a solar panel into your belly: it is a free recharge, courtesy of the sun. Eating conventional meat and vegetables produced on remote farms that use fertilizer and large machines is somewhere in the middle: the energy entering your belly is part sun, part petroleum: fossil food.

The energetic burden of a particular kind of food is closely, and sometimes confusingly, tied to its methods of production and distribution. Michael Pollan estimates that one of the least efficient food choices available at the supermarket is actually an organic vegetable. Lettuce grown in industrial-scale organic farms in California contain about eighty calories per pound; when transported to the East Coast markets, that lettuce requires about fifty-seven calories of fossil fuel energy per one calorie of food (167). Multiply 57 by 80, and you get nearly 4,600 petro-calories input to each plastic box of organic lettuce. Americans use nearly three times the energy that developing nations use to produce the same amount of food—that discrepancy in energy is fuel-based (Johnson 39). Clearly, eating is an agricultural and ethical act, and its complexities are not resolved by organic or vegetarian strictures: organic food uses more land than conventional, and it requires high-impact animal manure for fertilizer.

One reason this chapter is an interlude rather than a portion of the biomass or fossil-fuel sections is because our food is almost always a blend of sunlight and hydrocarbon. Think back on the last meal you ate that was grown or gathered by hand using no chemical fertilizer or pesticide, no machine transportation, and that was either raw or cooked over a wood fire (or one of those solar stoves distributed in developing countries). Anything bought from a store is disqualified, since even the local organic whole foods were brought in on trucks. Most cooked foods use the fossil-fuel calories that provide electricity or natural gas. Remember it? That meal represents pure biomass subsistence. I expect many of us have never eaten a single such meal, and the rest of us would appeal to garden tomatoes or their youthful forage of wild summer berries.

Literature about food often helps reduce its excessive connotative-ness down to basic relations between berry and body. These relations linger with the senses, taste and aroma, color and texture, and they help explicate how all the complex intimacy of eating can be clarified when it is paired with energy. Wild berries and nuts are a trope in literature of primordial satisfaction, since hunting and gathering is inherent to our evolution. In "August," Mary Oliver imagines her body and soul as a "thick paw of my life darting among / the black bells the leaves; there is / this happy tongue" (ll. 12–14). By synecdoche she reduces her human self to its animal portion, the paw and tongue, utterly satisfied with the wild harvest that is at once ritual and basic subsistence.

In "Fall," Wendell Berry gathers wild cherries like a latter-day Adam, enjoying "paradisal fruits that taste of no man's / sweat" (ll. 2–3). Though they represent the spontaneous bounty of nature, unsalted by human sweat and tears, they have a wild bitter flavor that requires patience from the palate. Berry concludes, "Reach up, pull down the laden branch, and eat; / When you have learned their bitterness, they taste sweet" (ll. 4–7). The sweetness operates on levels figurative and literal, but at its base the sweetness is our evolved preference for things that have calories, and the bitterness our evolved distrust of botanical toxins. When food is self-produced, stewardship cultivates pleasure. In Li-Young Lee's "From Blossoms," the fruit plays synecdoche for the idyllic landscape that fosters it: "To carry within us an orchard, to eat / not only the skin, but the shade, / not only the sugar, but the days, to hold / the fruit in our hands, adore it, then bite into / the round jubilance of peach" (ll. 12–16). The sweetness, in promising to prolong life, is a form of love between the animal and his habitat. Sugar love reaches back into its origins in the coevolutionary landscape, and the fertilizing, pollinating animal becomes a loyal steward by connecting seeds with the means of its production—sunlight, water, and time. John Keats invokes sugar as the emotive energy of joy, available to him "whose strenuous tongue / Can burst Joy's grape against his palate fine" (ll. 27–8). For Keats, these joyous sugars contrast the bitter sensations of melancholy, and result in the bittersweet ruminations of a life well-lived. Energy is life, and life is that higher-order flow of perception and reaction that unfurls in an unbroken series of instants.

Fruit tastes of sunlight-made-sugar, but butter is a more complex delectation. Butter odes abound, perhaps because butter's production is evocative at every stage. From the pastoral graze to the swollen udder to the nubile milkmaid to the strenuous churn master, butter has a series of connotations before it even gets to the table to adorn bread, to golden poultry, to mortar confections. Seamus Heaney calls it "coagulated sunlight" in his poem "Churning-Day," a close reading of the physical labor required to produce butter traditionally (l. 24). Following Heaney's lead, in "Ode to Butter" Linton Hopkins praises butter's material appeal, noting how the qualities of the pasture directly translate into the flavor of each batch, in effect flashing a biodiverse landscape upon the taster. Hopkins writes, "Cool and spreadable I taste your season, / bright, fat and herbal in spring and summer when / fed on clover and fresh grass / in the winter you taste of hay and grain" (ll. 28–31). Butter is as close as we get to being ungulates at graze in the fields, mildly noting the varied taste of each mouthful. Theriophily— admiration of animals for their serene presence in the present tense—emerges in butter odes as a compelling point of identification and sympathy between the cow and her caretaker. Hopkins concludes that butter's "very nature may be good for the fabric of our brain"—a nod not only to revised views of cholesterol science, but also to the figurative link between our fatty-brain imaginations and the network of life that conspires to bring butter into being (l. 76). The fabric of our brain is strengthened, literally and metaphorically, by its subsistence upon coagulated sunlight.

This unbroken gustatory chain from sunlight to season to forage to churn to table is such stuff as idylls are made of but the links can also be inconvenient for dairy farmers. In Thomas Hardy's *Tess of the D'Urbervilles*, for example, Farmer Craik's hard work is ruined by garlic in his grazing field. An entire batch of butter has a garlic twang, rather than the sweetness of clover and lavender. This is not an haute cuisine fuss; it is a matter of economic survival for a dairy farmer whose butter supplies the whole town. Craik sends his hands to root out the pungent invaders: "With eyes fixed upon the ground they crept slowly across a strip of the field, returning a little further down in such a manner that, when they should have finished, not a single inch of the pasture but would have fallen under the eye of some one of them. It was a most tedious business, not more than half a dozen shoots of garlic being discoverable in the whole field; yet such was the herb's pungency that probably one bite of it by one cow had been sufficient to season the whole dairy's produce for the day" (178).

In the twenty-first century, our palates are not trained to discern among the cow's selections because industrial butter is standardized, "quality-controlled." The cost of living in a world without garlic-twang butter is losing that leap of imagination back to the field and the sunlight that brought the butter into being. Heaney, Hopkins, and Hardy knew that butter did not come from the cold little box in the grocery store. In a poem more aligned with our times, Les Murray's "The Butter Factory" details the unappealing modern-day origins of that most bucolic spread: "It was built of things that must not mix: / paint, cream and water, fire and dusty oil" (ll. 1–2). The factory is fueled by biomass, an energy source that exacts hard labor: "The cordwood / our fathers cut for the furnace stood in walls / like the sleeper-stacks of a continental railway" (ll. 4–6). Food-making work is no longer the direct exchange between a provisioned landscape and the active harvester; in the industrial system it is triangulated through the factory outfit that demands its own fuel. And thus, along with grass and sunlight, timber, coal, and petroleum become present-day butter.

Sometimes food labor is pleasure turned to pain. Bringing in a ripe harvest requires prolonged and often manic days of work—calories exerted for calories secured. Robert Frost is haunted by the aural qualities of apples in "After Apple-Picking," and appetite turns to disgust as the body suffers in the aftermath of its labor: "I keep hearing from the cellar bin / The rumbling sound / Of load on load of apples coming in. / For I have had too much / Of apple-picking: I am overtired / Of the great harvest I myself desired" (ll. 24–9). Unlike Wendell Berry's Adamic pluck of the cherry, in Frost's poem the body suffers in its toil to preserve itself, a lapsarian economy between exerting and refueling. The setting is an orchard, where the harvest represents market value in a cash economy, rather than subsistence. The laborer's exhaustion comes from harvesting apples not only for himself but for the entire region—he runs an energy deficit. Even the apples "I myself desired" turn nauseous in their overabundance, a reworking of the pleasurable cornucopia.

Farm labor supplying a region with fruits is a form of civic service exacted upon a sore body, but the individual collecting wild fruits pays for his selfishness with a sore conscience. In "Nutting," the boy Wordsworth finds himself "rich beyond the wealth of kings" after he wrecks a wild hazelnut bower to collect its fruits for himself (l. 50). But this energetic wealth is attenuated by the "sense of pain" and guilt he feels from the "spirit in the woods" that seems to look in on the ruins of his incursion (l. 51, 55). In the unrestrained assault of "Blackberry-Picking," Seamus Heaney imagines a murderous scene in which "big dark blobs burned / Like a plate of eyes. Our hands were peppered / With thorn pricks, our palms sticky as Bluebeard's" (ll. 14–16). The coveted, visceral pleasures of these unwieldy sugar sacks turn to pain when Heaney loses them to spoilage. Nature deploys fungus and bacteria to reclaim the energy that humans squander by over-harvesting.

Perhaps most intriguing about these wild-food poems is their attention to guilt, a surprising emotion considering the light ecological footprint these particular food energies impose. Though they represent the least impact, they require close attention to the landscapes that provide them, so the gatherer is more aware of his actions' consequences. Wordsworth's "Nutting" is a microcosm: the actions of one boy in a bower extrapolate out to the violence of the entire human species against nature, which, Wordsworth observes, "deformed and sullied, / Patiently gave up [its] quiet being" (ll. 46–7). Stewardship does not work without qualms and moments of violation—these poignancies strengthen the ethic of care.

With most industrial-era food, sheer ignorance erodes individual responsibility. Methods of production are carefully hidden from the consumer, but literature has a knack for turning up the dirty underside of food making to act as an investigative report. The pastoral idyll is a suspicious nostalgia to readers who are aware of modern industrial methods, as Michael Pollan shows in his essay on "The Supermarket Pastoral" in *The Omnivore's Dilemma*. Corporate marketers strategically deploy pastoral imagery—quaint red barns and cud-chewing cows—to sell products that can be highly processed and full of artificial ingredients. We pay a premium for the bucolic narrative that we desire as the story of our food.

Upton Sinclair's *The Jungle* is a book-length study of the ways in which humans can oppress each other in the industrial food system. Laboring immigrants are preyed upon by a rabble of corrupt, greedy, and pitiless officials in Chicago's Packingtown. Their life conditions are luridly revolting:

> All day long the blazing midsummer sun beat down upon that square mile of abominations: upon tens of thousands of cattle crowded into pens whose wooden floors stank and steamed contagion; upon bare, blistering, cinder-strewn railroad tracks, and huge blocks of dingy meat factories, whose labyrinthine passages defied a breath of fresh air to penetrate them; and there were not merely rivers of hot blood, and carloads of moist flesh, and rendering vats and soap caldrons, glue factories and fertilizer tanks, that smelt like the craters of hell—there were also tons of garbage festering in the sun, and the greasy laundry of the workers

hung out to dry, and dining rooms littered with food and black with flies, and toilet rooms that were open sewers. (284)

Sinclair cleverly flips the appeal of food over to revulsion for the disgusting conditions of its production. The fetish gaze of his realist prose plunges into the gross materiality of factory work on such an unnaturally large scale. Poisoned rats are tidbits in the sausages. Workers that fall into rendering vats are canned with the lard. Rats, cows, and human workers share the same fate—literally to be eviscerated and eaten by the unsuspecting consumer. Sinclair characteristically appeals to unions and socialism as an anodyne to heal this corrupt, miserable food-making machine that literally eats up the bodies that labor to keep it running.

His novel's disgust factor supports a literary propaganda that did capture the public's attention. By direct consequence of *The Jungle*, the Meat Inspection Act and the Food and Drug Act were passed in 1906. These were anthropocentric measures; concern for animal welfare in the form of the Humane Slaughter Act would not emerge for another half-century (Spiegel xvii). Sinclair's novel is perhaps the most famous jeremiad against the industrial food system, enabled further by its publication at a time when industry and food were just beginning to fuse. The public's innocence was itself a tasty feast for the literary rhetorician.

More than a century later, our food system is more entrenched in the economics of biggering. Provisional improvements in worker, landscape, and animal rights acknowledged, we still operate within a system that looks a lot like Sinclair's Packingtown except it is even more sprawling and fossil-energy intensive. Billeh Nickerson's volume *McPoems* delves into his experiences as a McDonald's employee. Absurdity is on the value menu: in nearly every poem Nickerson cooks the absurd using measures of pathos and comedy. In "Diet," minimum-wage overwork causes him consume "two cheeseburgers, a large fries, and a big cola" every day—"a ritual ... to induce comfort and satisfaction" (ll. 3, 6, 7). The nutrition of the food and his compulsion to eat the same meal are hardly conscious, until a coworker "points out you've eaten / twenty eight cheeseburgers, fourteen large fries, and a kitchen sink's worth of cola"—the gross arithmetic of accumulation for the fast food addict in the spirit of *Super Size Me* (ll. 8–10). Nickerson snaps out of it: "You make a mental note / to pack a sandwich for lunch tomorrow, / eat an apple, possibly some celery" (ll. 11–13).

Many of the poems profile customers whose addiction to the high-density, low-quality fat, salt, and sugar-laced industrial food has become a defining feature of their lives. Nickerson can pick between customers who will order diet versus regular cola, and the woman who orders the salad with two packets of oily dressing. He writes of a bulimic woman who parks in the same location with her meal, leaving the vomit for a swarm of trained birds, and of the man who cruises the drive-thru four times in a day, from breakfast to dessert, and "pretends he doesn't know you, / you don't know him" (ll. 7–8). The drive-thru is the genius invention that united the exuberant car culture of 1950s America with the growing demand for cheap and easy mass-produced food, a culinary conjunction enabled by fossil fuels. Another drive-thru character is the "Fish-Filletmobile" filled with migrant workers: "In a heavy accent someone orders / fourteen fish burgers, parks the van for a while / then drives through again, / orders fourteen more" (ll. 5–8). These workers may have legal reasons to remain anonymous but, more broadly, anonymity is a powerful ally in shame culture. Nickerson's poems toe the line between a poet's sharp perceptions and the generalities of (usually bad and sad) behavior among his customers, who could be anybody.

It is the nature of the food that causes the behavior. Its nutritional profile causes addiction. Its culinary engineering sparks sensory appeal, as in "Alchemy": "Even you still marvel / at how quickly / everything transforms. / One moment it's frozen, / the next moment golden" (ll. 1–5). Its cheapness means that it is treated like garbage: pickles smacked on ceilings, soft-serve cones doubling as unicorn horns, ketchup that ensanguines the parking lot. Nobody pays attention to its origins or methods of production, distribution, preparation. When a chicken truck overturns in front of the restaurant, sprinkling the property with "zonked / out chickens," a man runs over one with his car (ll. 5–6). He screams through the drive-thru window that "he didn't take his family out to dinner to kill something. That made you think a long time" (ll. 8–10). Apparently, a McChicken sandwich is so processed that it no longer represents the body of an animal.

Fast food is culture as much as food, and Nickerson's volume teases out the rituals that keep this machine running: minimum wage, minimal quality, maximum volume, cheap energy, and the evolutionary pleasure buttons of taste that delude our industrial-era brains into tasting survival, rather than obesity and diabetes, in fat, sugar, and salt. For some customers, particularly the poor and immobile, the food does represent a form of survival since it is the only food offered in North America's "food deserts" (Nickerson is Canadian). He tells the story of an elderly man who orders a hundred cheeseburgers: "He intends to freeze them, they'll get him through the winter, no need for pesky walks on cold days, no danger of slipping and breaking a hip. *100 cheeseburgers will keep me going for a little while longer, at least. I don't need much*" (ll. 7–11).

The fundamental message beneath the volume's absurdity is that the fossil food Nickerson serves is a cultural failure of care and responsibility, for the workers, the consumers, and the animals and agricultural lands that produce it. The shadowy corporate beneficiaries emerge only in the trite mantra from the training manual, which Nickerson chooses as his epigraph: "The recipe for success must always include quality, service, cleanliness, and value. These are your most important ingredients" (ll. 1–3). Nickerson deconstructs this folderol by dividing his poems into sections that lampoon each of the four pillars of fast food.

Entrenched as we are in this industrial food system, what steps might we take to disengage? Gerald Stern is a guerilla gardener amid the factories of Easton, Pennsylvania on the Delaware River. In "Planting Strawberries," his pleasure comes not only from fruit sugars, but also from a revolutionary postapocalyptic vision of agriculture reclaiming trammeled places:

What I like best is having a garden this close to
the factories and stores of Easton.
It is like carrying a knife in my pocket!
It is like kissing in the streets!
I would like to convert all the new spaces
Back into trees and rocks.
I would like to turn the earth up after the bulldozers
have gone and plant corn and tomatoes.
I would like to guard our new property—with helmets and dogs.
I would like us to feed ourselves in the middle of their civilization.

(ll. 15–24)

Agriculture has entered a time in which traditional, low-energy, down-scale methods of production are having a renaissance in the face of the tragicomic wastage of the industrial food system. With local farmers' markets and urban agricultural systems on the rise, we are beginning to kiss in the streets again. As with all other aspects of life, industrialism injected fossil fuel energy into agriculture, massively ramping up scale and yield while eliding issues of environmental damage, ethical failure, and consumer health.

As developed countries reconsider the costs of industrial scale, energy dwells at the heart of the matter. We evolved as opportunistic omnivores, subsisting entirely on biomass energy. With industrialization, our omnivory came to include fossil fuels: we became petrovores. As we look for solutions in the twenty-first century, the sprawling infrastructure that enables the mass production of low-quality processed food at heaping profits for corporations, all enabled by hydrocarbons, will come under more scrutiny as an unsustainable system designed to feed Mammon. If Gerald Stern gets his way, our strawberries will taste like the day they were picked. But we may have to stoop to pick them, our muscles working the sugars of the sun.

Part Three

Primary Energy

8

Renewables: The Elements Move

The fundamental source of earth's energy, our sun, has infused biomass energy into the planet since the first chlorophyll cells evolved in cyanobacteria a few billion years ago. This evolutionary innovation fabricates chemical energy, sugars, from the energy in photons. As we have seen, almost all the food and fuel we have ever used has come as a consequence of solar rays driving photosynthesis. Geothermal energy is the rare exception. We usually gobble solar energy through plant vectors, both recently alive as biomass and long dead as fossil fuels. But this googolplex of solar voltage has only been captured directly to perform work for a few decades, since the first solar panels were launched into space to power satellites in the late 1950s. Viewed from the economics of fuel, the sun is an energy paradox: a hemisphere of energy constantly hits the earth, but it is dispersed and sporadic in many cloudy regions, so the energetic infinity of solar infusion dwindles down to a scattering of solar panels powering a small fraction of our machines.

The discrepancy is less evident with wind and water, the elemental engines put in motion by the sun. Since antiquity, humans have built mills to yoke the power in their motion. Watermills were especially important prime movers in preindustrial civilization, where a constant supply of energy was required to grind and press and saw the raw harvests into consumable goods like flour, oil, and lumber. The manufacturing centers of early modern Europe tapped into primary energy by clustering around regions of moving water, which must be counted among the essential natural resources of the continent (Cipolla 93). The limiting factor on the expansion of preindustrial European economies was a severe lack of energy, seen especially in the overconsumption of biomass like timber and charcoal.

Medieval and early modern Europeans built mills wherever coursing water allowed, thus supplementing the restrictive energy economy, as the river performed the work that other societies extracted from slave labor and from animals (powered by their food, biomass). While fossil fuels could be transported where they were needed, thus freeing up the geography of industrial production and distribution, rivers could not be moved on locomotives. Recall William Wordsworth's relief that coal had saved the beautiful Lake District from the incursions of industries powered by moving water. The vogue of channel building in late eighteenth-century England pushed the limits of moving water as a form of transportation, but its development was soon aborted by the coal-fired railway. As fossil-driven engines took over the energy economy in the nineteenth century, the age-old infrastructure of river mills was mostly dismantled or abandoned.

When it is not being tapped, primary energy disappears as a fuel source, reverting to dynamic nature itself—the sun, wind, and water. Once coal, then petroleum, then natural gas became available to power engines, primary energy performed a disappearing act. With the exception of the massive hydroelectric dams built for flood control and electricity in the American New Deal era of the twentieth century, and analogous modern installations especially in China and India, primary energy has remained hidden in plain sight for most of the Industrial Revolution. Only in the last few decades have we intensively researched the technologies of capturing it again. The most conscious use of fossil fuels would be to build a renewable energy infrastructure that will capture primary energy for generations to come. Consider coal, petroleum, and especially natural gas as platform fuels that allow us to manufacture the green infrastructure of wind turbines, tidal stream generators, hydroelectric dams, and solar panels. We can use the world's most rich and problematic fuels no longer as a material supporting addiction to a wasteful high-energy global economy, but instead as a segue to a more sustainable regimen of renewable primary energy sources.

Because of its diachronic, retro/futuristic history, the literature of primary energy tends either to look back with nostalgia on simpler, quieter, cleaner times of yore, or to peer forward into the dazzling potentials of carbon-free energy captured from the natural elements and atoms. As this chapter will show, however, primary energy has a pessimistic, even apocalyptic, life in literature as well. Hydroelectric dams devastate riparian landscapes, drowning entire canyon systems in the American West and enabling huge population increases in Sun Belt cities that could not exist without the dams' water and electricity. And that most controversial source of primary energy, splitting the nucleus of the atom, came online when the world was still reeling from the atomic ending of the Second World War and gearing up for the long trudge of the Cold War. The twentieth century's drama of nuclear fission for electricity and the bomb inspired some of the strongest literary reactions to fuel ever written.

I Breeze

Other than water mills and their more modest cousin, windmills, one other source of primary energy had an indelible effect on the growth of preindustrial civilization: wind over water, captured in the full sail of the ship. Societies built upon navigable waterways with access to the ocean have been historically dominant—from the Egyptians, Greeks, and Phoenicians of antiquity, to the Spanish and Portuguese in the age of discovery, to Britannia ruling the waves in the age of empire. Wind on water made a global economy possible. Wind on water fueled the exchange of raw natural resources for finished goods, and it also supplied labor to the shores of production, most infamously in the Middle

Figure 8.1 Rembrandt van Rijn, *The Mill*, 1645. Image courtesy of the National Gallery of Art. Public domain.

Passage between Africa and the Americas, which consigned tens of thousands of Africans to their deaths on the high seas, and millions more to slavery.

Sailing ships opened the world first to intercontinental discovery, then to colonization, and later to the exchange of goods across the globe. Sailing ships were our first power tool, lending a sense of how we might live as superhumans liberated from the meager means of our animal muscles. By the apogee of the sail, the eighteenth century, the most efficient clipper ships were able to capture more than two hundred times the power from wind in exchange for the muscle power that operated them (200–1 is still meagre compared to the 500–1 exchange of a coal miner) (Nye 18). With superhuman power comes all the avarice and cruelty, all the wanderlust and gumption, that have marked expansionist societies since the ancient Egyptians first lashed planks together into a hull and set sail on the Red Sea five thousand years ago.

Aeolus is the god of the wind and its keeper. In some Greek myths, he is the son of the ocean god Poseidon, as the wind and the sea had a close relation in the Greek mind. The second oldest piece of Western literature—Homer's *Odyssey*— gives Aeolus an essential role in deploying wind energy in order to assist the world-weary Odysseus as he meanders the seas traveling back to Ithaca. Living a contented life among his large and incestuous divine family, Homer's version of the wind god does not embody the restless power of his device. Aeolus feasts Odysseus and his men on his eponymous floating island and, hearing their plight, grants them the best present the globe has to offer: a steady West Wind directly back to their native shore. This being an epic, Aeolus complicates the benediction by also giving them a Pandora-like windbag, literally "a wallet, made of the hide of an ox of nine seasons old, which he let flay, and therein he bound the ways of all the noisy winds" (10.120). Aeolus secures his treacherous present "fast in the hold of the ship with a shining silver thong, that not the faintest breath might escape," and he sends the ship's crew along their full-sailed way west, to Ithaca (10.120).

For ten days Odysseus holds three sheets to the wind (quite soberly), trusting no other hand to sail. This work tires him out, and as they sight Ithaca, with its welcoming beacon burning brightly onshore, the hero falls into a luckless sleep. His sailors take this opportunity to satisfy their curiosity about the gold that may lurk in Odysseus's purse, going below decks and piercing the windbag: "And the violent blast seized my men, and bare them towards the high seas weeping, away from their own country ... the vessels were driven by the evil storm-wind back to the isle Aeolian, and my company made moan" (10.122). Petty avarice unleashes a cosmic fury. Odysseus receives no comfort or understanding from

Aeolus, who observes that a man so little under control of his hands is "the most reprobate of living men. Far be it from me to help or to further that man whom the blessed gods abhor!" (10.122). Odysseus must be cursed, if the winds of north, south, and east conspire against his course true west.

In *Ulysses*, James Joyce adapts this "epic fail" to modern times in his Aeolus chapter, set in the Freeman newspaper offices. Joyce evokes the windbag in its more familiar metaphorical sense by having the characters mock the inflated rhetoric of various speechifiers. Where the classical heroes wrestle with god-sent zephyrs, Bloom and his compatriots in palaver flail about in murky metonymy and chaining chiasmus.

In Virgil's *Aenead*, Aeolus is a god who keeps the winds enchained in a deep cave, acting as both jailor and counsellor. Aeolus whispers to them, "softening their passions, tempering their rage: if not, / they'd surely carry off seas and lands and the highest heavens, / with them, in rapid flight, and sweep them through the air" (1.57–9). Virgil's metaphor for the wind's contained power comes from his audience's closest acquaintance with force: the horse. Aeolus knows how to "tighten or slacken the reigns" to deploy wind energy to his advantage, and he does so when Juno promises to him her beautiful daughter Deiopea in exchange for a devastating storm that would destroy the sailing fleet of the Trojan Aeneas (1.63). The furious, choreographed storm leaves Aeneas wishing he had perished on the shores of Illium with Hector, rather than suffer the tumult of water, sand, and wind that scuttles his fleet. Happily for Aeneas, the gods have monstrous egos and Neptune is angry that his ocean is disordered by Aeolus's winds, which he orders back to their cave in the lonely mountains. Neptune paternally soothes his seas and Aeneus's bedraggled fleet seeks desultory winds to tack toward the Libyan coast, where they may "stretch their brine-caked bodies on the shore" (1.173).

These early invocations of the wind indicate an extension beyond a material resource into the realm of luck and fate, ministers of the gods. Mortals contend with an entity that cannot be controlled in the way that land resources may. The wind is all-powerful and ineffable; it is held only in the small measure of a sail's breadth; like rain, it is summoned by supernatural intervention rather than scientific study of the material world. The wind behaves as a physical force that might help or hinder mortals, but it simultaneously represents an augury. To do work in this preindustrial world, to move heavy objects through space, requires a benediction of wind rather than a methodical use of an available resource. It requires prayer and supplication. It also requires knowledge, technique, and moxie to navigate an indifferent wind by tacking and harboring, and tolerate mystery, vagary, and misfortune. This mindset of being laid open to the fortunes of an equivocal cosmos is quite foreign to the industrial world, with its ready stores of reliably intense fuel-feeding hungry engines that are under human control. In the wind world, progress is not inevitable; it is fortuitous. When fortune blows the other way, one must sit and wait, and perhaps even contemplate the futility of ambition in a world of self-cancelling forces. The sailor beneath the sail understands the limits of his autonomy; the sailor beneath the ship's smoke stack often does not.

In many early literary uses, the wind is an engine of divine will. In effect, the wind acts as a physical indicator of something inscrutable—it makes the hidden will of the gods known to man. Even if unjust, capricious, and cruel, this blustery force unites the supernatural with the natural and maps out the human place in the cosmos. The un-reason of divine will translates directly into our rational perception of physical phenomena, supplanting agency and justice with fact and effect. The gods must be angry: my ship faces a headwind. If not rational, this dynamic is at least perceptible and was highly influential in literary history. The topos of wind literature lingers in purely metaphorical grounds, where the poet likens his condition to the bewildered mariner's plight. Early modern writers were fond of using the chagrin of the luckless sailor to capture the emotional pains of the lost, disfavored, and unloved poet. Sir Thomas Wyatt senior, that famous lamenter who modified the Italian sonnet for English use in the sixteenth century, shows how the wind can fill an extended metaphor. Here is his sonnet titled by its first line:

> My galley, chargèd with forgetfulness,
> Thorough sharp seas in winter nights doth pass
> 'Tween rock and rock; and eke mine en'my, alas,
> That is my lord, steereth with cruelness;
> And every owre a thought in readiness,
> As though that death were light in such a case.
> An endless wind doth tear the sail apace
> Of forced sighs and trusty fearfulness.
> A rain of tears, a cloud of dark disdain,
> Hath done the weared cords great hinderance;
> Wreathèd with error and eke with ignorance.
> The stars be hid that led me to this pain;
> Drownèd is Reason that should me comfort,
> And I remain despairing of the port.

(ll. 1–14)

The galley, a ship most often propelled by muscle power using oars, is here a vehicle that carries the poet's grief across deep fathoms worried by uncertain winds. Emotion imposes all the violence of a god-stirred angry sea. Wyatt crosses hopeless love ("mine en'my" and "my lord") with sea nature, translating the adventure crisis into an existential one: the literal voyage exists only to draw out a metaphorical one. Likewise, the "endless wind" that "doth tear the sail apace" is a colorful way to imagine emotional adversity and bewilderment, as the poet suffers from stellar disorientation ("the stars" were a Renaissance conceit for a lady's fine eyes). This is a poem about heartache, but its entire imaginative structure is built upon wind power: specifically our inability to make earth's primary energy work for us.

Shakespeare's more famous sonnet 116 ("Let me not to the marriage of true minds") describes a very different journey: Love "is an ever-fixèd mark / That looks on tempests and is never shaken; / It is the star to every wandering bark, / Whose worth's unknown, although his height be taken" (ll. 5–8). In the two sonnets, the same elements assemble to opposite effect: for Shakespeare, love defies an adverse wind; love is a North Star that keeps the voyager on course. These poets were employing the tools available to their time in history and their culture. After the British navy defeated the Spanish Armada in 1588 and was known as the world's mightiest, maritime pride was prominent in British culture. However, the 130 ships in the Armada were less devastated by the English fireship attacks than they were by a fierce north Atlantic storm that roared for two weeks and wrecked many ships on the Scottish and Irish coasts (Konstam 13).

Elizabeth I gave thanks to a Protestant God for delivering England from the Catholic Spaniards, and had a medal minted with the words *Flavit Jehovah et Dissipati Sunt*—"God blew and they were scattered." The so-called Protestant Wind made a legend of the Virgin Queen and strengthened the Protestant cause across Europe. The action of the elements was taken to represent divine will; luck was interpreted as destiny. Some modern scholars attribute this vicious weather pattern in 1588 to the accumulation of polar ice in the North Sea caused by the Little Ice Age (Fagan xvi). Far from supernatural, the Protestant Wind was a tantrum thrown by an unbalanced climate, the kind of anomaly that is familiar to our intemperate times.

Still, giving winds have a magical quality about them, as if summoned from good spirits and operating on the level of incantation rather than meteorology. In the middle of that pirate-infested, sea-shanty century, the eighteenth, Olaudah Equiano's narrative of the African Prince includes an account of the Middle Passage. Having never seen a ship before, Equiano asks the white sailors how the vessel works. He is told about the mast and the ropes, but Equiano continues to believe that the bad men put a magic spell in the water to propel the abducted across the sea. How could nature's winds be the sole power for such a diabolical journey?

Samuel Taylor Coleridge's "Rime of the Ancient Mariner" brings these supernatural qualities to the wind, and his reckoning with the elements results in a parable of virtue and transgression in the natural world. In such a famously weird, haunted poem, it is easy to forget that ships like the mariner's, leaving from imperialist British ports, had mandates for trade or exploration. The fanciful winds, in league with other uncontrollable forces like weather and disease, were read as a text that indicated some divine opinion on the enterprise, just as the Greeks had read the winds. Favorable winds literally carried the voyage across the seas and figuratively enclosed the ship in a bubble of supernatural favor. In Coleridge's superstitious poem, as soon as the mariner commits the inexplicable sin of killing the albatross, the sailors start their close reading of the wind to see what judgment might ensue. At first "the good south wind still blew behind," escorting the ship from cold Antarctic waters, through the Torrid Zone toward the equator (l. 87). This good omen merely tricks the sailors into league with the sinful mariner, and once they all are of a bird-slaying mind, "Down dropt the breeze, the sails drops down, / 'Twas sad as sad could be; / And we did speak only to break / The silence of the sea!" (ll. 107–10).

The ocean seems to die around them; the sun desiccates the living: "Water water everywhere, / Nor any drop to drink" (ll. 121–2). They enter a thirsty purgatory, trapped on a windless ocean so still it seems petrified. The only remaining element is the damning sun hammering down upon their misery. Meanwhile, personified "Death" sails on a windless ocean; she pulls broadside their becalmed vessel and parts souls from bodies. But the original sinner, the mariner, must stay on board until he is absolved for his act of wanton destruction, after he blesses the sea snakes. Once the sea snakes receive that blessing, they seem to breathe life into listless sails: "And soon I heard the roaring wind: / It did not come anear; / But with its sound it shook the sails, / That were so thin and sere" (ll. 309–12). Here, Coleridge chooses to move entirely into the supernatural element while leaving the literal wind behind. The roaring wind dissipates and no breeze ensues, yet "Slowly and smoothly went the ship, / Moved onward from beneath" (ll. 375–6). By disinvesting the metaphor of the wind from its connotation of divine benediction, Coleridge makes the ship's progress suspect and the mariner's fate equivocal, as the zephyrs whisper "The man hath penance done, / And penance more will do" (ll. 408–9).

The riddle of the wind extends through the poem, evading any neat explanations while cultivating a sense that the elements in this poem are choreographed by equivocal fates. The sails, thin like dead leaves, are no longer the powerhouses of a weighty British enterprise, they are tattered ghosts flitting above the decks of a cursed vessel. Estranged from their design and purpose, the sails signify a complex human failure not only to control the elements, but even to understand the exchanges between act and fate in a super/natural world. Coleridge tempts us with Homeric clichés of the benedictory wind, but all that remains in the end is an empty vessel of a man uttering the tag line, "He prayeth best, who loveth best / All things both great and small" (614–5). Liberated from the moral machinations of judgment and favor, the wind in the "Rime of the Ancient Mariner" signifies a secret life of nature that defies human understanding, a Romantic zephyr indeed. Within Coleridge's lifetime, the wind would become superfluous to the world of trade and travel. Humans no longer needed divine favor to drive across the seas; we needed only coal.

As the sail made way for the steam-driven propeller, many of the aesthetic qualities of being at sea changed. The expertise required to cross safely lapsed, replaced by a new set of engineering skills far removed from traditional wind aptitudes in tacking, beating, reaching, and running. Because the source of energy moved from external to internal power, the crew's attention shifted from the natural elements to the machine working within its hull. Stellar and coastal orientation gave way to instrumentation. The ship's range became not a question of skill, season, and prevailing winds, but a calculus of fuel efficiency. Like many of the analogous changes in transportation brought by fossil fuels, abandoning wind power implied leaving behind an acute perception of the cosmos in favor of an infatuation with internal engine works. This involves more than spending time inside industrial machines (trains, cars, ships) instead of hauling yards under the twinkling stars.

It is truly an ontological shift in which our existence changes from a relationship between humankind and nature to a relationship between self and mechanism. The world shrinks down, from a global cosmos of tumult and calm in which nature is a character in its own right, to an egocentric negotiation between humans and our gizmos. Natural conditions become merely a stage for human/machine action, a device for heightening the drama in a storm or dressing up romance in a sunset. Sailing ships sink while battering external conditions; steamships often sink because their massive internal energy causes ruinous momentum. The essence of failure or success depends upon the design of the contraption and the competence of the people trained to operate it. The scuttling of the Titanic was due more to its own coal-fired velocity ordered by a captain thirsty for fame, than an innocuous iceberg floating on calm seas.

Now that two centuries of coal and diesel ships have greatly reorganized the global economy and shrunk our sense of space-time, would the seas be enhanced, perhaps even re-enchanted, by a return to wind power? In the heyday of steam-powered shipping, Wilkie Collins's 1871 novella *Miss or Mrs?* uses the distinction between steam and sail to draw a contrast between an ambitious, corrupt businessman and a languorous young man on board a small sailboat off the English coast. The vessel has been becalmed for two days, a situation that distinguishes the pragmatic Turlington from the romantic Linzie. The former demands a prompt return to London trade; the latter is content to wander at the whim of wind and tide. Turlington snarls, "Next season I'll have the vessel fitted with engines. I hate this!" (8). Linzie replies, "Think of the filthy coals, and the infernal vibration, and leave your beautiful schooner as she is. We are out for a holiday. Let the wind and the sea take a holiday too" (8). The characters of Turlington and Linzie are aligned with their fuel preference: Turlington represents avarice, impatience, and egotism; Linzie is passive, perceptive, and fitful.

Frequently in Victorian novels, the heroine selects among opposing male types, and that is the case here. Natalie Graybrooke is a fifteen-year-old heiress who, in a curious example of medical counsel, has been prescribed a sea voyage to counteract her rapid growth into womanhood—as though the sea-by-sail preserves a fresh innocence and feminine lines. Though the effect on her body is uncertain, the voyage affects her mind as "a delicious languor in her eyes, and an utter inability to devote herself to anything which took the shape of a serious occupation" (10). Far from preserving her innocence, Miss Graybrooke discovers her sexuality during the trip, having a secret affair with Linzie in the close quarters of the hold. Their encounters suggestively take place in the storeroom, where "tea, sugar, and spices were at her back, a side of bacon swung over her head, and a net full of lemons dangled before her face" (21). Whatever figurative fruits they manage to pluck in youthful haste, the exotic goods in the store room were certainly imported using coal power.

In their conversations, the lovers banter about the vagaries of the sail-borne voyage. Natalie eulogizes their special time at sea sequestered from the bustle of the working world: "How I shall miss the wash of the water at my ear, and the ring of the bell on deck, when I am awake at night on land! No interest there in how the wind blows, or how the sails are set. No asking your way of the sun, when you are lost, with a little brass instrument and a morsel of pencil and paper. No delightful wandering wherever the wind takes you, without the worry of planning beforehand where you are to go. Oh, how I shall miss the dear changeable inconstant sea! And how sorry I am I'm not a man and a sailor!" (11). Natalie's wish to be unsexed is clearly more of a rhetorical flourish than an actual desire to make a living by the sail, as she has an heiress's idealized notion of rigorous work. Before long, the breeze blows again and the motley crew are harbor-bound. Turlington reaches the shore by a rowboat (biomass powered), crowing "Pull, you lazy beggars! ... Pull for your lives!" (29). The rest of the tale is landlocked; as Natalie predicted, the wind and sail are soon forgotten, the romance interrupted, and wasteful getting and spending resume.

This vignette of a sailing voyage suggests several of the exchanges that took place as steam replaced primary and biomass energy in industrial-era transportation. Collins identifies a critical moment in the history of sailing when wind-blown ships became idle pleasure vessels for the wealthy, and steamships operating along strict timetables took control of global trade. Being becalmed, then, implied pleasure and relaxation rather than a divine curse or bad luck. Old money could afford to linger under the idle sail; new money needed steam to chuff back to port. Released from the work of the world, sailing became a hobby of the leisure class and, with it, close reading of nature-at-sea shifted from a required skill of the mariner to a pastime of elite pleasure seekers. Nature itself ceased to work; production settled on a human-centered nexus of exchange driven by coal, a worker that could be deployed at will.

Antiquated skills like sailing and fox hunting are aristocratic trademarks because they represent time spent not making money. In contemporary times, "hobbies" like sewing, woodworking, and gardening represent former occupations that have been pushed toward obsolete by a fossil-fueled global industrial economy that provides clothes, furniture, and food in exchange for money. Still, the artisan in each of us yearns for the creative and physical outlet that used to represent skills necessary for survival. Obsolescence is the bailiwick of the privileged. Collins marks the generation where the shift from the working wind to the pleasure breeze was a paradigm change in transportation that distinguished between the classes.

If schooners, sloops, and galleons are impossibly antiquated, what would a futuristic sailing ship look like? Paolo Bacigalupi asks the question in his novel *Ship Breaker* (2010), which is set in a dystopian American Gulf Coast very late in the fossil-fuel era. The ship breakers are children who crawl through the guts of dilapidated oil tankers to reclaim their valuable scrap metals, particularly copper, for reuse. The hero, a boy named Nailer, is one of them. He describes a scene that is already familiar amid today's general decay of industrial infrastructure:

> From the height of the tanker's deck, Bright Sands Beach stretched into the distance, a tarred expanse of sand and puddled seawater, littered with the savaged bodies of other oil tankers and freighters ... flayed and stripped, showing rusty iron girder bones. Hulls lay like chunks of cleavered fish: a conning tower here, a crew quarters there, the prow of an oil tanker pointing straight up to the sky.... wherever the huge ships lay, scavenge gangs like Nailer's swarmed like flies. Chewing away at iron meat and bones. Dragging the old world's flesh up the beach to the scrap weighing scales and the recycling smelters that burned 24–7. (6–7)

The United States has become an underdeveloped, lagging region oppressed by pollution and rising seas. Like vultures, the poor gather scraps to sell to the rich who live overseas. Nailer and his crew are postindustrial chimney sweepers, conscripted by their small size and poverty to the most heinous work available. He will soon crash out of an internal duct into a pool of petroleum deep in the ship's hold, a valuable find in this era of fuel scarcity, but a potentially fatal one in context. Before he finds a way out, Nailer thinks, "I'm gonna drown in goddamn money" (26). After he luckily slips out of a side hatch, "a tear in the tanker's hull still spewed oil, marking where the ship had vomited him into open air. Black streams of crude traced down the ship's hide, running in slick rivulets"—a passage that pulses with disgust at the sickness of the old petro-world (35–6).

There is no old world without a new world, and poor boy Nailer observes that new world navigating far off the despoiled coast of Louisiana:

> Nailer's eyes followed a clipper ship as it sliced across the waters, sleek and fast and completely out of reach.... a clipper with high-altitude parasails extended far above it—sails that Bapi said could reach the jet streams and yank a clipper across a smooth ocean at more than fifty-five knots, flying above the waves on hydrofoils, tearing through foam and salt water, slicing across the ocean into Africa and India, to the Europeans and the Nipponese. (7–8)

Notable in this passage is the conjunction of high technology with an elemental, free, primary energy source majestically liberated from the cumbersome, polluted world order of petroleum. The clipper ship uses high-altitude wind power. It is both simple and technological, clean and complex, enabled rather than delimited by its reliance on the wind. No longer are sailors beholden to the caprice of Aeolus; a power source as constant as the sun exists in the jet streams, a global conveyor belt of energy. But the technology required to capture a jet stream at forty thousand feet is out of reach. The wind is free; the dingus that taps it is dear. What is a poor boy to do but dream? Nailer imagines being on deck, "hurtling across the waves, blasting through spray … The ships whispered promises of speed and salt air and open horizons" (46). We are furnished a closer look when a hurricane wrecks a clipper on an offshore island and Nailer attempts rescue:

> Even destroyed, it was a beautiful thing, utterly unlike the rusting iron and steel hulks they tore apart every day. The clipper was big, a ship used for fast transit and freight on the Pole Run, over the tip of the world to Russia and Nippon. … Its hydrofoils were retracted, but with the carbon-polymer hull shattered, Nailer could see into its workings: the huge gears that extended the foils, the complex hydraulics and precision electronic systems. (79)

The comparison between the "rusting dinosaurs … great wallowing brutes leaking their grime and toxins into the water" and this "machine angels had built" is superficially based on a contrast between material and technology (80). But the fundamental distinction is energy, chemical versus kinetic. The machine follows the aesthetic of its fuel source: tankers are designed around heavy metal tanks to hold oil; clippers run clean over the surface of the waters, defiant of gravity and inertia, and are drawn on by an unseen power. Their technique is the work of angels.

Ship Breaker imagines a technology that has been a point of speculation for nearly two hundred years, since John Etzler published his utopian *The Paradise within the Reach of all Men, without Labor, by Powers of Nature and Machinery*, in 1833. Though he imagined tapping high-altitude wind power, his accomplishments were more modest. Etzler holds an 1842 patent for a peculiar fan-like sail powered by an on-deck windmill that allows only one man to manage the rigging.

The ship is also equipped with a basic hydrofoil technology that reduces hull drag in the water. Etzler's design did not remake shipping; after all, in the 1840s he was competing with coal, the miracle fuel of industrialization. But his ideas have been rekindled by generations ever since, and currently there is an airborne wind energy consortium that organizes ideas and technology in the sector (however, the consortium has not updated its website since its inaugural 2010 conference). The total kinetic energy in high-altitude winds is estimated to be around a hundred times the energy that all human activity currently consumes. Working prototypes of kite-like flying energy generators have been built, but their long-term viability, economics, and environmental impact require more study (Lipow). The future of the technology is still in the realm of *Ship Breaker*'s science fiction, but the race is on, with venture capital behind it.

If fuel sourcing signifies shifts in human ontology, the sailship represents the limits of control humans can exert over natural conditions. This limited agency leads to religious and superstitious rituals meant to appease gods and nature, a set of behaviors largely eliminated by the onset of the fossil-fueled engine ship. However, Bacigalupi's high-tech clipper ship returns agency to human technology while tapping the unlimited resource of the wind. If this science fiction vision is applied to future commerce, it will reconcile human ambition with agency, again defeating the preindustrial ontology of a human fate determined by a capricious natural world.

II Flow

Rivers and streams laid a web of kinetic energy across the preindustrial landscape. The best sites for water mills lay in steep river valleys, where water fell sharply through space and the wheel sipped the force of gravity from the flow.

Figure 8.2 Joseph Mallord William Turner and Robert Dunkarton, *Water Mill*, 1812. Image courtesy of the National Gallery of Art. Public domain.

But population centers rarely clustered near these high-energy, non-navigable streams. Increasing that kinetic energy from its natural potential, subject to seasonal fluctuation, anomalous weather, and inconvenient location, involved some of the most complex engineering ever seen before the engine. Cities like London and Boston were built with access to the sea and sailship in mind but, lacking a steep river valley, inland water power required engineering to make low-gradient rivers turn mill wheels.

One famous example is the spectacle of the fountains at Versailles, where an enormous contraption known as the machine at Marly was constructed in the seventeenth century to power the show. The machine used fourteen enormous paddle wheels in the flow of the Seine to power a series of pumps that moved river water six hundred meters uphill to the Louvreciennes aqueduct. Once that potential energy of nearly five hundred vertical feet was gained from the kinetic energy of the river, the water naturally flowed back downhill to supply the chateau and gardens with a pressurized flow. The system daily consumed the same volume of water as the whole city of Paris. The fountains of Apollo, Bacchus, and Latona spouted appreciatively, their leisure quite at odds with the labor of sixty workers who kept the imbroglio of machinery operating. King Louis XIV oversaw the seven-year construction, and in June 1684 the machine began its labors, which continued clear through the French Revolution and Napoleonic wars, until 1817 (Pendery).

An outstanding symbol of royal decadence, the machine at Marly was a beast of engineering prescient of the monstrous hydroelectric dams of the twentieth century, which convert that same kinetic energy of river flow into electricity, a more fungible form of energy than a waterwheel. Flowing water has more than symbolic and aesthetic powers, however. The availability of water power in the steep valleys of New England was one crucial resource that drove the northern states to a highly productive industrial economy while the American south, a region of shallower gradients, was still clinging to slave muscle power to produce its agricultural goods (Nye 55).

In England, hydropower changed over the centuries from wild meandering rivers to a network of dammed, locked, and channeled water courses designed to draw the most power most consistently from sporadic flows. Lord Byron teasingly compared the engineered millraces to their wild predecessors using a love metaphor, in his satirical epic *Don Juan*:

> There was Miss Millpond, smooth as summer's sea,
> That usual paragon, an only daughter,

Who seem'd the cream of equanimity

    Till skimm'd—and then there was some milk and water,

With a slight shade of blue too, it might be,

    Beneath the surface; but what did it matter?

Love's riotous, but marriage should have quiet,

And being consumptive, live on a milk diet. (Canto 15. 321–8)

If young love is the wild spring-flood river, marriage is the mirror-still millpond, tranquil on the surface but with depths hard to plumb. Byron was one to sip the cream and turn away the meager bluish remains, but his image of the domesticated woman/pond preserves the latent power in the waterworks, the "slight shade of blue" indicating both mood and mystery. If not a flood rush, Miss Millpond embodies weight in motion, and therefore energy.

Akin to the nostalgia for horses and country roads that set in with the locomotive, the Victorians took stock of their old hydraulic infrastructure as the competing coal-fired steam engine chuffed onto the scene. The old technology was nothing to sniff at. Water mills provided major gains in energy: a small, four-horsepower grist mill produced the energy of a hundred working men without needing to be fed or housed, but only requiring a millrace with adequate flow (Nye 22). In 1835 the early Victorian Andrew Ure looked back at the virtues of waterpower in comparison with the coal revolution. He surveys a watermill at the village of Belper that is "driven altogether by eighteen magnificent water-wheels, possessing the power of 600 horses ... As no steam engines are employed, this manufacturing village has quite the picturesque air of an Italian scene, with its river, overhanging woods, and distant range of hills" (343–4). That the old technology could possess at once such power and ease, uniting the best of English productivity with Italian aesthetics, is why Ure—who spends most of his book detailing the wonders of coal-driven steam—could look with admiration on the prior alternative to the new, noisy, dangerous, dirty mill driver.

The steam engine aggressively boils water and uses its pressure to drive the piston rod, but the fluvial mill is a passive mechanism with wheels dipped in the active river flow. The relationships among water, mechanism, and the energy source are diametrically opposed in these two technologies. Moreover, the greater the degree of intensification, by pressure, tension, or velocity within a machine's physics, the more extreme the violence inherent in malfunction. Steam engines could blow up when over-pressurized, with extreme carnage such as the hundred passengers instantly killed when the steamship *Oronoko* exploded in 1838 (Nye 84).

As fossil fuels quickly replaced primary energy in the nineteenth century, the philosophy of fuel and force received deeper scrutiny. It was easy to purchase a pile of coal and claim it as your private store of energy, but who actually owned the primary power that came from rivers and winds? In an 1854 lecture called "The Interaction of Natural Forces," Hermann von Helmholtz noted that "the possessor of a mill claims the gravity of the descending rivulet, or the living force of the moving wind, as his possession," an egocentric position that is hard to defend in claims court (501). The palpable liquid of the water itself is only the vector of energy; the source is gravity making the water descend through space. Elaborate networks of mill ponds and races were engineered not only to contain flow for future use, but also to supplement the natural gravity of the river's course. When water was diverted upstream, the medium of energy was tapped, but the potential energy of gravity is elusive, immaterial, unclaimable.

Helmholtz implies that to assume possession of such an ineffable and universal force betrays an egotism mismatched to the essence of natural primary power (Ketabgian 118). Such egocentric claims were much better suited to the coming age of fossil-fired industry and wealth privatization: with coal mines and oil fields developers can make legal claims, not with immaterial gravity and sunshine. The entire industrial–consumerist revolution of the last two centuries has reinforced the individual ego of the manufacturer, the entrepreneur, the consumer, and the lawyers who preserve the private ownership of property, particularly fossil fuels. Private profits are the foundation of capitalist enterprise, and this element of legal ownership further distinguishes elemental-fuel ontology from the fossil-fuel ontology that succeeded it.

Dinah Craik's novel *John Halifax, Gentleman* is a moralistic Victorian novel, one of a flock of nostalgic tomes written mid-century with an eye cast back to the arcadia of a preindustrial countryside, including a rags-to-riches hero who evolves into an elective gentleman through good hard work. It was hugely popular in its time, the type of book that concerned parents would foist on their worldly children, but now it remains largely unread because it seems stultifying to modern readers. Nonetheless it retains value in regard to what it says about the fuel wars of the nineteenth century. Central to the story is Halifax's decision to convert his cloth mill from water power to fossil fuel by replacing the millworks with a steam engine. This choice is more than a practical exchange of fuel and force; it is emblematic of a mid-century view that the steam engine represents the upwardly mobile man and his inherent virtues of autonomy, discipline, hard work, and measured emotions (Ketabgian 109). The Victorian industrial hero contrasts with the bygone Romantic heroes, whose throes of passion suggested energy in its native forms: great waves, floods, gales, and thunderstorms. Halifax outmaneuvers Luddites and Lords to navigate his mill into Victorian prosperity, while still managing to preserve the natural beauty of the region—at least there are no Dickensian soot flakes falling on rickety orphans.

The Avon river, the one that Shakespeare's Stratford is built upon, powers an old flour mill that predates Halifax's cloth mill: "Here [was] a narrow, sluggish stream, but capable … of being roused into fierceness and foam. Now it slipped on quietly enough, contenting itself with turning a flour-mill hard by, the lazy whirr of which made a sleepy, incessant monotone which I was fond of hearing" (44–5). Halifax recalls the strangely quiet tumult of the old works as a peace made exciting by the power driving the millworks: "The mill was a queer, musty, silent place, especially the machinery room, the sole flooring of which was the dark, dangerous stream" (107). The biotic mesh of this quiet countryside is complemented by the mill, whose aural qualities only enhance the soporific contentment of an idyll. The peace relies on motion, not stillness. The watermill continues its latent and lazy productivity that contrasts the active, violent clamour of the coming steam engine.

Halifax does not modernize in order to downsize his workforce and replace hands with automation, as Luddites of the eighteen-teens warned. He is forced from stream to steam because the Avon's flow is squelched by a self-indulgent aristocrat, Lord Luxmore, who reroutes river water for his manor's fountains. His appropriation has an immediate environmental effect. Before, Halifax's children "spent whole mornings in the mill meadows. Through them the stream on which the machinery depended was led by various contrivances, checked or increased in its flow, making small ponds, or locks, or waterfalls. We used to stay for hours listening to its murmur, to the sharp, strange cry of the swans that were kept there, and the twitter of the water-hen to her young among the reeds" (325). After, the idyll is killed by mechanism: "Only twice a week the great water-wheel, the delight of our little Edwin as it had once been of his father, might be seen slowly turning; and the water-courses along the meadows, with their mechanically-forced channels, and their pretty sham cataracts, were almost always low or dry. It ceased to be a pleasure to walk in the green hollow, between the two grassy hills, which heretofore Muriel and I had liked even better than the Flat. Now she missed the noise of the water—the cry of the water-hens—the stirring of the reeds" (332). The engineering of the millrace blends seamlessly with the pastoral wetland: water itself seems to delight in the "pretty sham cataract." Drawing off the flow of water depresses the children and the waterfowl as the wetland reverts to a more mundane dry grass community. Water is life, and the water-starved meadow is dying.

Halifax has no legal recourse, especially not against a lord named Luxmore, so he seeks redress in the new technology, a secret project involving a "mass of iron, and the curiously-shaped brickwork" (333). The novel pivots from the idyllic stability of generations-old waterworks that supplied a number of local industries, to the new order of steam-driven mills, each with its own autonomous power station. With the river a diminished thing, the next generation will presumably fix upon the great engine in the new mill rather than on the subtle patterns of the river's millrace flow and its community of species. Power lies in the boiler, a scalding, hostile, enthralling dragon; it is no longer associated with the lazy river or the built-in mechanics that regulated its flow and captured its power.

Rather than proceeding from nature, to the new generation fuel and force will appear an entirely man-made enterprise, a genius architecture of brick, iron, and glass, of railway cars piled with man-mined coal, of workers assisting and regulating the machinery that now does the moving and making. Craik does not further eulogize this decoupling of energy from nature and the marriage of fossil fuel with industrial civilization. She concerns herself with Halifax's alternatives. He uses coal and steam to defy gravity and rise above the old-world destiny that would leave him, along with the lowering river, in the mud (Ketabgian 113). The rest is history, and this one is a comedy: marriages, wealth, prosperity, enabled by ready-fire coal. The cloth mill outdoes its former hydro-powered self and becomes a symbol of Victorian victory.

Lord Luxmore's fountains run full, though he dies a hell-bent bankrupt. The quiet fuel of the *fleuve* becomes an anachronism, the food of nostalgia, as the wind powered the pleasure yacht in Collins's novel. The river powers the appearance of prosperity—fountains to festoon a bankrupt estate—rather than performing the real work of the prosperous. It is superfluous, pastoral "bling": fountains to complement sun-struck lawns and topiaries lined by gravel walks, in the manner of Versailles. Aristocratic estates use energy and space for pleasure and to demonstrate wealth. They designate acreage for sprawling lawns and gardens rather than using it to grow useful biomass that feeds people and working animals, or to capture the energy directly with windmills and water-mills.

This freedom to cache energy for enjoyment is a feature of developed societies with wealth inequality. To own the resource and use it only as a figure of thought or for personal, privileged enjoyment shows a degree of power over the landscape and other people. In *Pride and Prejudice*, when Elizabeth Bennet admires Mr. Darcy's estate Pemberley, her first observations are essentially tallies of standing reserves of energy. Her carriage winds through "a beautiful wood, stretching over a wide extent" and once they pass all this proud biomass they see the manor itself, "and in front, a stream of some natural importance was swelled into greater, but without any artificial appearance" (185). Pemberley's river supports gentleman's leisure fishing (as its woods are a deer park), but it does nothing so practical as turn a wheel to power the estate's energy needs. The machination of swelling an important stream into an even greater force of nature serves purely aesthetic demands. We soon learn that Mr. Darcy is not so snooty that he would exclude Elizabeth's working-class uncle from enjoying his trout stream. Uncle Gardiner is invited into the club for a day; he basks in Pemberley's cultured energy excesses and we might imagine he returns to the coal scuttle of London with a renewed sense of the advantages of wealth. Money creates pleasure bubbles of effervescent primary energy.

Contemporary to Craik's novel is the more famous work *The Mill on the Floss*, by George Eliot, her second novel published in 1860 after *Adam Bede*. *The Mill on the Floss* is notably poor in descriptions of the riverine scene; probably the best evocation is offered on the second page, where "the rush of the water, and the booming of the mill, bring a dreamy deafness, which seems to heighten the peacefulness of the scene" (8). A bit later, the heroine Maggie Tulliver helps us in the fossil-fuel age imagine what kind of work was accomplished by the falling water: "The resolute din, the unresting motion of the great stones, giving her a dim delicious awe as at the presence of an uncontrollable force—the meal forever pouring, pouring—the fine white powder softening all surfaces, and making the very spider-nets look like faery lace-work—the sweet pure scent of the meal—all helped to make Maggie feel that the mill was a little world apart from her outside everyday life" (26). The flour mill grinds meal for the entire region, so the water's work is essential to the welfare of surrounding families; it lands on their table as bread. The "presence of an uncontrollable force" carries with it an augury: this remembrance of things past is poignant because the old mill's works are soon to be swept away in a sixty-year flood.

The portions of the tragedy that deal with the water mill take two forms: litigious and diluvian. Before the flood, the miller Mr. Tulliver is consumed by a lawsuit against men upriver who threaten to dam and irrigate his mill's river, the Ripple, a tributary to the larger Floss. Being a miller is a ticklish role, since Tulliver has to constantly attend to the river's water levels and flow, grain prices, the millworks themselves, and his family of motley personalities. With such responsibilities, the town imagines him to be a man of some wealth whose demands will be met, but it is not so. Tulliver's legal fate is reminiscent of Jarndyce versus Jarndyce in Dickens's *Bleak House*: he becomes embroiled in a tortuous battle incomprehensible to its claimant and, ultimately, after a lot of lucre has been lost to court costs, he loses. Puzzling over the tangle of legal positions relating to water claims brings out Tulliver's character:

> I don't know what I shall be forced to; but I know what I shall force him to, with his dykes and erigations, if there's any law to be brought to bear o' the right side.... . Water's a very particular thing—you can't pick it up with a pitchfork. That's why it's been nuts to Old Harry and the lawyers. It's plain enough what's the rights and wrongs of water, if you look at it straightforrard; for a river's a river, and if you've got a mill, you must have water to turn it; and it's no use telling me, Pivart's erigation and nonsense won't stop my wheel; I know what belongs to water better than that. Talk to me o' what th' engineers say! I say it's common sense, as Pivart's dykes must do me an injury. (130)

Tulliver's commonsense confidence that "water was water," however, does not account for sovereign rights to the river's flow from any point upstream (132). He loses to the detested lawyer Wakem in court, a battlefield ill-selected for an uneducated miller, and Tulliver further degrades himself by assaulting Wakem on his horse. Tulliver's death comes hard upon his defeat.

The Tulliver children Tom and Maggie are reduced to poverty as their family's assets come up for auction and their golden days of youth become a bitter memory. The rest of the novel explores their relationship with each other and with Maggie's suitors in the context of a fallen family that has lost its ancestral mill. Father Tulliver's pride, superstition, daydreaming, and rampant speculation on how things ought to have been fix him in a nostalgic mode that contrasts with the industrial worldliness of the modern men competing to take over the mill. Mr. Deane serves as old Tulliver's foil as he speaks to young Tom Tulliver about the future of the mill: "The world goes on at a smarter pace now than it did when I was a young fellow ... forty years ago ... the looms went slowish, and the fashions didn't alter quite so fast ... it's this steam, you see, that has made the difference: it drives on every wheel at a double pace, and the wheel of fortune along with 'em" (320). Deane and his rivals all wish to see the river as something, if not superfluous, at least merely supplemental to the new prime mover that spins the modern world's wheels: steam. Father Tulliver is a sacrifice to the industrial age, a sad fool whose own personal failings hasten him to desuetude.

But the now-superfluous river forces the novel's climax. Eliot's use of this classic literary topos—the cleansing, annihilating flood—brings primary energy back into the action. Tulliver had been dreading an existence with too little water drawn off by men upstream. But the ruin of his mill is realized instead by the stunning force of flood water in proportions remembered only by the elderly residents who witnessed a similar event sixty years before. Maggie the Innocent lacks respect for the river's awesome power: "There was hope that the rain would abate by the morrow; threatenings of a worse kind, from sudden thaws after falls of snow, had often passed off in the experience of the younger ones; and at the very worst, the banks would be sure to break lower down the river when the tide came in with violence, and so the waters would be carried off, without causing more than temporary inconvenience, and losses that would be felt only by the poorer sort, whom charity would relieve" (414). As with many natural disasters, the interval between disastrous events is often just a bit longer than an individual's power to recollect—poignant at a time when so-called hundred-year floods seem to happen at a decadal or even faster interval. This flood will be felt by more than "the poorer sort" who live on low land.

Maggie spends the last hours of her life attempting to save others. She snatches her beloved brother Tom out of one death in the mill house only to join him in another, drawn into the torrent by "some wooden machinery [that] had just given way on one of the wharves" (421). Though the passage is ambiguous, the siblings may literally have drowned under the weight of their family's own mill wheels. For a novel that has spent hundreds of pages carefully delineating the lives of the young Tullivers as they endure life's trials, this flood is a device that sweeps away the preceding narrative structure and leaves the ending bare. The biblical flood, which had been incorporated into geological theories of catastrophe, had in Eliot's time been critiqued by the gradualists Lyell and Darwin (Otis 236). Eliot uses a geological aesthetics that includes both perspectives: layers of narrative accretion gradually thicken her text, then she sweeps a catastrophe over that settled shoal of river sand.

By replacing the river's gradual ebb with a disastrous flow, *The Mill on the Floss* records the human toll of expecting what is naturally a variable, pulsing system to behave as if it were a controlled, mechanical one. The hubris and naivety of this position come from those early images of the mill as "dreamy" and "peaceful." Primary energy lulls us into complacency; we are its passive recipients. This golden mood causes us to lose sight of the coming calamities of too-low and too-high water. No wonder that every victor in these Victorian races rides a steam engine.

The complex of millraces that fueled preindustrial production seems modest in relation to industrial-era hydropower. Though considered a "green" energy source, modern hydropower involves massive dams that span the world's most powerful rivers. Natural dams created by landslides, log jams, and aquatic animals predate humanity. The first artificial dams, dating back to 3000 BC in Jordan, were built to assure constant water supply for agricultural people. The vocation of dams changed very little until the Industrial Revolution, when they were built on a mass scale to supply power to large corn and cloth mills (Turpin 10). The hydroelectric dam is an industrial-era re-versioning of the old fashioned millworks, which had a network of ponds, races, dams, and wheels. The water turbine used by a dam naturally evolved from the water wheel of the river mill, with one key difference: it could produce power from the river and send it over long distances to urban and manufacturing centers where electricity was needed (Turpin 39). Cities and mills no longer needed to cluster along the river banks to use river power.

Modern dam makers accept the trade-offs inherent in their work: for electricity, flood control, and stable reservoirs, we sacrifice the flooded river valleys, their history, and resident species—such as salmon, which can only travel on free-flowing rivers—and the livelihoods of people who rely on a naturally pulsing river system. Long before Glen Canyon in the United States was drowned by its eponymous dam in the middle of the twentieth century (more on that later), the Elan Valley scheme in Wales submerged Cwm Elan in 1893, a valley that the poet Percy Shelley had described in 1811 as a scene with rocks "piled on to each other to tremendous heights, rivers formed into cataracts by their projections, and valleys clothed with woods, [all of which] present an appearance of enchantment" (qtd. in Turpin 76). Queen Victoria approved the series of dams in the Elan valley in 1892 to support a disease-free water supply to industrial Birmingham.

One last sacrifice, the most abstract, is the wild river itself as a symbol of natural flux and creative variability. Natural systems and their species have evolved in variable conditions that present endless challenges to survival. Dams effectively squelch seasonal extremes in water level and manage flow to benefit consumers of water and electricity. In some locales, the cost–benefit of hydroelectric dams has shifted back against them, as in the case of the 1913-built Elwah Dam in Washington State. The value of the spawning salmon that swim upriver (but not over concrete monoliths) and the free-flowing riparian community now exceeds that of the electricity the dam produced for nearly a century. Despite the trend of decommissioning century-old dams across the United States, at the global level dam building for electricity is on the rise. Dams supply more than 16 percent of global electricity demands with an annual increase of about 3 percent. Globally, in just the last decade, power generation capacity at hydroelectric dams has increased by nearly a trillion kilowatt hours to 3.5 trillion (Moller).

Industrial-era dams excite bipolar emotions. Their sheer scale and clean concrete lines set against the uneven forms of rocks and foliage make them an engineer's sublime. Architects recognized the aesthetic potential of dams set in wild country. In 1945 Charles Holden, a designer for the London Underground, said that the engineers building dams have an aesthetic responsibility: "Dams are invariably situated in the heart of wild and mountainous country, and it is in your power to give accent to the grandeur of the landscape or alternatively to cast a blight over the whole neighbourhood by thoughtless or callous indifference" (qtd. in Turpin 116). Many of the great dams in Europe and America were commissioned during the mid-twentieth century, when the towers, spillways, and stately arches along dam faces were designed in enormous proportions, replacing the flouncy ornamentations of the nineteenth century (Turpin 131). The engineers of Hoover Dam brought in an architect to design its forms in an art deco style, notable in the massive intake towers that thrust out of Lake Mead on the upstream side.

Modern literature often portrays dams as the result of ill-advised and sinister human intervention: controlling the wild river is tantamount to turning it into a machine servile to humans. The hydroelectric dam has been powerfully deployed in the western United States to supply electricity and reserve water for dry Sun Belt cities like Phoenix, Las Vegas, and Los Angeles. These cities would not support their present populations without a massive infrastructure of dams and aqueducts installed in the twentieth century. However, for years before the proposed project got underway, an elderly John Muir and his Sierra Club opposed flooding the Hetch Hetchy Valley with the Tuolumne River in Yosemite National Park, part of the O'Shaughnessy dam project. This dam supplies water and electricity to San Francisco, but for Muir it was a destructive commercial scheme. Muir celebrated the valley's rock masses and majestic waterfalls, its species diversity and the calm it brought to the soul of the explorer:

> Hetch Hetchy Valley, far from being a plain, common, rock-bound meadow, as many who have not seen it seem to suppose, is a grand landscape garden, one of Nature's rarest and most precious mountain temples. As in Yosemite, the sublime rocks of its walls seem to glow with life, whether leaning back in repose or standing erect in thoughtful attitudes, giving welcome to storms and calms alike, their brows in the sky, their feet set in the groves and gay flowery meadows, while birds, bees, and butterflies help the river and waterfalls to stir all the air into music—things frail and fleeting and types of permanence meeting

Figure 8.3 Dolly Salazar, *Hoover Dam Intake Tower*, 2010. Image courtesy of Wikimedia Commons. Public domain.

here and blending, just as they do in Yosemite, to draw her lovers into close and confiding communion with her....

These temple destroyers, devotees of ravaging commercialism, seem to have a perfect contempt for Nature, and, instead of lifting their eyes to the God of the mountains, lift them to the Almighty Dollar.

Dam Hetch Hetchy! As well dam for water-tanks the people's cathedrals and churches, for no holier temple has ever been consecrated by the heart of man. (813–4, 817)

Muir appeals to religious and sanctified images to convey a sense of how Hetch Hetchy is not a blank wilderness, but a natural treasure analogous to humanity's greatest devotional structures. Part of the horror of these modern dams is due to their sublime scale, both in what is gained and what is sacrificed. As a younger man, Muir built dams on his own farm stream in Wisconsin that aided irrigation and flood control (Turpin 233). But the essence of the peril in the twentieth century became the superhuman scale of industrial construction aided by fossil fuels. The self-perpetuating logic of these projects excused the costs by extolling the benefits—a platform for future development that would in turn require more water, and more power.

Despite the Sierra Club campaign, the O'Shaughnessy Dam was completed in 1923, and Hetch Hetchy was flooded. To preservationists, it is a tragedy. To developers, a triumphal comedy: San Francisco enjoys some of the cleanest municipal drinking water in the world, and the electrical generation from the dam's two power stations is about 1.6 billion kilowatt hours per year (San Francisco Department of the Environment). Dismantling the system to recover the Hetch Hetchy would foist the burden of energy onto another source, likely a fossil one. Ambivalence about dams continues to surface in interdisciplinary, interpolitical debate in which the sacrifices of last century continue to haunt and reward a consuming society.

To Zane Grey, the prolific pulp fiction writer of the American west, the human-centered glory of such outlandish public works projects like Hoover Dam was "an idea born of the progress of the world, as heroic and colossal as the inventive genius of engineers could conjure, as staggering and vain as the hopes of the builders of the pyramids, an idea that mounted irresistibly despite the mockery of an unconquerable nature—and it was to dam this ravaging river, to block and conserve its floods, to harness its incalculable power, to make it a tool of man" (3). Grey's hyperbole in the novel *Boulder Dam* (another name for Hoover Dam) is tailored to the scale of modern ambition, and he swings repeatedly between awe for the scale of the work and suspicion of the "colossal egotism" that would promote such impositions on wilderness. Much of the story reflects a patently conventional 1960s America, with a football-playing hero whose *bildungsroman* parallels the story of the dam that enthralls him; he outgrows his snobby ex-girlfriend; he brawls with a "dirty Red" who tries to blow up the construction site; he recovers his kidnapped girlfriend in a lightly wounding gun battle with "white slavers"; he earns respect from the chief engineer as both athlete and intellect (211, 226). Gee-whiz observations pour out of the mouths of dam workers as the concrete amasses; under the cruel desert sun they dream of the tourist paradise they will find on the shores of Lake Mead, to be created by the dam.

One of the more interesting aspects of this novel is Grey's interest in the theme of sublime nature versus stupendous engineering, the "romance and tragedy of the man-made scheme to improve on nature" (58). His ambivalence is clear. From scenes of reverence for mighty man his prose turns to disdain: "Piles of debris, lumber, iron, slopes of talus, huge muddy holes in the river bed, trucks winding along the threads of roads. All of which appeared to make a hideous monstrosity imposed upon the sublimity of nature by thousands of workmen who resembled crawling ants" (141). Grey turns to an apocalyptic romantic vision of the futility of human ambition to outlast nature. Oft-cited spectacular quantities—4.5 million cubic yards of cement, nineteen million pounds of reinforced steel, a new reservoir of 227 square miles—fail to spare human accomplishment from the rubble of deep time (57). Ruins have a distinct kind of beauty that invokes the forces of nature in deep time and the failures of civilization to hold back entropy, but one would not expect ruins to figure heavily in a hagiography of Hoover Dam.

The novel notably departs from its heroic narrative by considering ambition and futility as equally legitimate outcomes to massive infrastructure projects. Grey's final vision opposes the simple beauty of engineered form with the ragged beauty of deep time, agent of decay. The dam is completed with "a splendid macadamized road, with its white carved stone parapet" to accommodate sightseers. "Boulder Dam was beautiful in the extreme. It had sublimity, grace, translucent changing color, and perpetual melody and movement. But all this, and Lynn's elation, could not restrain his mind from envisioning the grand scene five hundred thousand years hence" (280–1). Nothing of the dam's shining concrete remains in the deep future, leaving only glaciers and floods, water moving powerfully through the unpeopled landscape. The Colorado will flow "dirty white, remorseless, and eternal" (282).

For wild-minded writers fond of a depopulated desert scene, dams are anathema, emblems of an industrial state that equates nature with natural resource. Grey's reverent chronicle of the rise coexists with his rogue fantasy of the fall. Edward Abbey, our laureate of ecological mayhem, only wants to tear down industrial erections. In the nonfiction work *Desert Solitaire*, he makes special plans to raft down the Colorado through Glen Canyon prior to its flooding. In a long, lyrical chapter that traverses 150 miles of the canyon, Abbey discovers "an Eden, a portion of the earth's original paradise" as he and his companion meditate on wilderness versus "syphilization" (152, 160). Marveling at countless side canyons they can only glimpse at the river's pace, Abbey describes "the walls rise up again, slick and monolithic, in color a blend of pink, buff, yellow, orange, overlaid in part with a glaze of 'desert varnish' (iron oxide) or streaked in certain places with vertical draperies of black organic stains, the residue from plant life beyond the rim and from the hanging gardens that flourish in the deep grottoes high on the walls. Some of these alcoves are like great amphitheaters, large as the Hollywood Bowl, big enough for God's own symphony orchestra" (161–2). From this chromatic splendor that seems to extend by synesthesia to a grand musical scale, Abbey foregrounds the action of his future ecoterrorist novel *The Monkey Wrench Gang* by imagining the president himself "ignite the loveliest explosion ever seen by man, reducing the great dam to a heap of rubble in the path of the river" (165). Abbey imagines thrilling new river rapids from the dam ruins, then dismisses the entire image as the product of "idle, foolish, futile daydreams" (165).

Picking up where the machine-breaking Luddites left off in early industrial England, the Monkey Wrench Gang specializes in the sabotage (*sabot*: wooden shoes; *-age*: damage done to machinery by sabots) of any machinery used to develop the deserts of the American southwest. As detailed in the fossil-fuels chapter, George Hayduke and his pals are most proficient at scuttling diesel machines like bulldozers, killing the industrial beast by a thousand cuts. The greatest fantasy of this motley troupe of eco-moralistic miscreants is to blow up the Glen Canyon Dam, "a plug, a block, a fat wedge" on the diminished Colorado River (2): a fatal stab to the heart dealt out by "a law higher" than the human laws that permit rampant development (27). Seldom Seen Smith goes down on his knees on Glen Canyon Bridge and looks skyward:

> Okay, God, I'm back... . It's me again, Smith, and I see you still ain't done nothing about this here dam. Now you know as well as me that if them goddamn Government men get this dam filled up with water it's gonna flood more canyons, suffocate more trees, drown more deer and generally ruin the neighborhood... . All we need here, God, is one little *pre*-cision earthquake. Just one surgical strike. You can do it right now, right this very second; me and George here won't mind, we'll go down with the bridge and all these innocent strangers come here from every state in the Union to admire this great work of man. How about it? (157–8)

God does not immediately respond, so Smith and Hayduke ponder their mortal options: TNT in the control room (too much security); explosive houseboats crashed into the dam (problems with ignition). Instead they get drunk at the local bar in Page, Arizona, and fail in their attempts to stir a bar brawl.

The gang sets its sights on more modest targets, but the great dam remains the central symbol of their discontent. The object itself, and everything altered to make way for it, hideously exceeds in cost the electricity that is produced by the dam. The contemplative Dr. Sarvis wonders at

> all this fantastic effort—giant machines, road networks, strip mines, conveyor belt, pipelines, slurry lines, loading towers, railway and electric train, hundred-million-dollar coal-burning power plant; ten thousand miles of high-tension towers and high-voltage power lines; the devastation of the landscape, the destruction of Indian homes and Indian grazing lands, ... all that ball-breaking labor and all that backbreaking expense and all that heartbreaking insult to land and sky and human heart, for what? All that for what? Why, to light the lamps of Phoenix suburbs not yet built, to run the air conditioners of San Diego and Los Angeles, to illuminate shopping-center parking lots at two in the morning, ... to keep alive that phosphorescent putrefying glory (all the glory there is left) called Down Town, Night Time, Wonderville, U.S.A. (173)

The source of the gang's anger is that all the destructive infrastructure listed in Abbey's ranting parataxis is installed to support superficial self-gratifying consumers who know nothing about the landscape sacrificed to provide them with electricity. The dam is a symbol of a meddling human society whose ideals privilege consumerism over conservation, industry over nature, creature comfort over wilderness.

Written in the early 1970s, Abbey's novel partakes of the protest counterculture that sprung up out of real environmental crises (petro-polluted Lake Erie caught fire in 1969, for example), and his radicalism inspired enduring eco-fringe collectives like Earth First! The Glen Canyon Institute is a group dedicated to dismantling the dam and restoring the submerged canyon to its previous ecosystem. Hoover Dam was built as a public works project in the 1930s before there were widespread organized environmental groups or environmental impact studies; the case is similar with Glen Canyon Dam, which was approved during the concrete-happy Eisenhower administration.

The thriller *Wet Desert* by Gary Hansen picks up where Abbey's fantasies come back to earth. Hansen conducts a thought experiment regarding the hydro-physical impacts of an ecoterrorist attack that destroys Glen Canyon Dam. Meditating on the melee to come, the terrorist visits the river before his work on the dam:

> Even here in the Grand Canyon, the river was shackled, bound like a prisoner, unable to show its full strength.... The concrete dams held it back. Sure the river ran a little stronger than yesterday. But that only meant the flow through the turbines at the Glen Canyon Dam, some hundred and seventy-five miles upstream, had been increased, most likely due to a "hot one" in Phoenix, when

the Arizonians cranked up the air conditioners. More electricity from the turbines meant more water downstream. It was as simple as that. The mighty Colorado River was a slave to man, caged and controlled. (10)

Clearly invoking an Abbey's ideology, the bomber notes how abstracted the river has become from the water and electrical resources that it provides to consumers. Society imposes no ethical connection between the plug in the suburban wall and the ruined 1,500 mile riparian ecosystem. To the activist's chagrin, places like Phoenix and Las Vegas seem to flout their excesses, eagerly slurping water for fancy fountains to create the appearance of oasis in what is truly a hostile and hardly habitable stretch of the desert. False paradises come at the cost of the real oases in lush grottoes along the lost Glen Canyon. Often with dam construction, the national benefits of water regulation and electricity trump the local sacrifices of settled land and wildlife habitat. Someone from the dam-managing Bureau of Reclamations eventually profiles the attacker as an environmentalist who seeks to restore the Colorado River Delta that used to flow into the Gulf of California in Mexico. The bomber wants to restore the dry delta to the great marsh that existed until the 1930s, and he succeeds, for a short while.

The novel's ending is prescient of the celebrated 2014 pulse flow of the Colorado all the way to the sea. That reunion of the river with its delta briefly reanimated the delta's ecology, and gave a sense of what landscapes might be recovered if just a fraction less water were used upstream in the desert states. Water is only becoming scarcer and more valuable, however. The debate about big mid-century American dams is a complex one that defies clear political polarities: boaters on Lake Powell (created by the flooding of Glen Canyon) are often environmentalists; dam breakers are frequently libertarians, like Abbey's monkey wrenchers, who litter the desert with beer cans and busted bulldozers.

Dams like Glen Canyon and Hoover are often cited as evidence of environmental sustainability. The dams have existed long enough so their negative environmental impacts have been dissolved in a re-formed landscape. In place of drowned Glen Canyon, Lake Powell is celebrated as a bastion of recreation, fresh water and flood mitigation. Downriver, Hoover Dam supplies electricity to 1.3 million people without releasing any greenhouse gasses. The National Park Service's history of the dam shows how endemic these alien structures have become: "It is an American icon, a monument to the ingenuity of the nation's engineers and the power of its machines. Hoover Dam is the symbol of an era when an urban, industrial America reveled in harnessing its natural resources" (NPS.gov). The horse in the harness, the Colorado River, is in an extended drought worse than any in 1,250 years. With the loss of water comes an uncertain future for the booming populations of desert states and the important agriculture of California's Imperial Valley, which supplies 15 percent of the nation's food (Wines). The Glen Canyon and Hoover dams are only as mighty as the river they tame, and the system is currently on a life support, with recycled sewage effluent, new siphoning tunnels, and plans for rationing in coming years.

The era of the big dam has passed in the United States. Since the 1990s, Americans have been in the era of decommissioning dams and restoring the watersheds to a preindustrial condition. Globally, however, building big dams is an ongoing development initiative. Their ability to irrigate dry land, regulate floods, and provide clean electricity is too tempting to resist in places like Nigeria, China, and Brazil. Hoover Dam was the largest in the world when it was built in the 1930s; now it is out of the top thirty for size, and does not rank in the top hundred for electricity generation (Turpin 234).

A dam built upon the rhetoric of hyperbole (the longest, the tallest, the most expensive, the most energy produced), the Three Gorges Dam across the mighty Yangtze River is a prime example of the trade-offs involving industrialized primary energy. Fully operational as of 2012, the Three Gorges Dam translates the Yangtze's flow into 22,500 megawatts of electricity, a flood of juice that supplies one-tenth of all of China's electricity needs that would otherwise be fulfilled by coal. The power both enriches the economy and reduces pollution in Shanghai, at the Yangtze River Delta. The dam also controls floods that overran residential and agricultural land about once a decade. To accomplish this feat, the government had to forcibly relocate over a million residents, sacrifice riverside archaeological sites dating back thousands of years, and flood the natural wonder of the sharp-peaked Three Gorges themselves—an ironic name for the dam. Most controversially, the dam would have been impossible to impose upon societies with greater individual freedom since so many villages and towns were drowned. But the communist government has an authority over its people that forces them to truckle to its agenda. At best, the Three Gorges project represents a low-carbon future economy in which gravity and water take over coal's tasks and push China toward sustainable productivity. China already has the plurality of the world's large dams: some twenty thousand of a global total of forty-five thousand stand within its borders (Turpin 10).

The literature involving dams considers the polarities that spring out of their existence. Engineers are genius, arrogant, naïve, and foolhardy. Environmentalists are gentle river stewards and/or ecoterrorists. Consumers are ignorant, callous, self-absorbed, curious, and playful. Dams are tourist marvels, monuments of the industrial sublime, and ugly clots in the earth's arteries. The old river system is a lost paradise, wilderness, monster, and wasteland. The new river system is a slave, weakling, tool, machine, and lifeblood to agriculture and development. These irreconcilable perspectives coexist in the complex life histories of the world's great rivers. Early hydropower built on water wheels was different in scale, but not in spirit: we have, for thousands of years, attempted to make the river serve our needs and prevent its ravages insofar as our technology allowed. The concrete megaliths of the twentieth century appeared to bring that dance between nature and culture to an end by imposing the human mandate. Now, less than a century later, the dams are silting up and the river is drawing down. A future chapter opens on the postindustrial scene, with the great dams reduced to concrete scars on the eternal canyon walls. Perhaps the history of dams teaches us as more about ourselves than about nature. Ecosystems continually morph to accommodate changed conditions. Human ecology must likewise evolve by accepting that our engineering monuments will crumble under the pressure of geological time.

III Star

The reputed last words of the British painter J. M. W. Turner were, "The Sun is God." Our star is the origin of nearly all earthly fuel. Sunlight animates the winds, tides, and the hydrocycle that drives precipitation and river flow. Sunlight excites chlorophyll cells in plants, allowing them to store chemical energy in the form of sugars. Photosynthesis creates biomass fuel and, with enough deep time, fossil fuel. Without the sun, earth would never have existed, let alone incubated and sustained the riot of life—evolving, diverse, ubiquitous—that makes earth unique in our knowledge. The sun is a chemical entity, a fury of hot plasma slowly running down on deep time scales irrelevant to human concerns. That chemical reaction is the energetic foundation of life. What if we actually witnessed its extinction?

That question has stimulated apocalyptic literature through the ages, as eclipses, periods of lower solar activity, sunspots, and volcanic eruptions with parasol effects have imposed climate changes on cowering human societies. In Renaissance conceits the sun is figured as the constant, strong (male) star opposed to the fickle (female) moon, but the sun, too, varies in its intensity and our planet annually orbits at varying distances, its hemispheres pitched toward the sun in summer and away in winter. In the Gaia metaphor, the goddess Earth "breathes" each year as deciduous leaves take up carbon in spring and exhale it in autumn, a breath in rhythm with the changeable sun. Vicissitude, even within annual patterns, is one reason life on earth is so varied and restless: abscission, migration, hibernation, and provisioning are all organismal behaviors evolved to deal with an inconstant sun. Seasons are profoundly important in designating organism relationships within a biotic community and with the abiotic elements like water, which changes from gas to liquid to solid depending on solar activity.

Seasons are also crucial to human identity, individual and cultural. Literary autumn lapses into metaphor as soon as the leaves begin to fall. As Shakespeare observed in sonnet 97 on the seasons as a metaphor for love, "How like a winter hath my absence been / From thee, the pleasure of the fleeting year! / What freezings have I felt, what dark days seen! / What old December's bareness everywhere!" (ll. 1–4). His absence from his love has been dark and cold, indeed, but the second quatrain reveals that he actually shivers in summertime. This implies, not uniquely, that his love is as the sun, and her (or his?) absence from his summer makes it subjectively a season of decline and death, even as the natural world mocks his agony with a "teeming Autumn, big with rich increase" (l. 6). Without his love on hand, these fecund swellings are "widowed wombs … orphans and unfather'd fruit," leaving the absent love to play the role of father and the speaker to play the widow (ll. 8, 10). Gender reversals or same-sex desire aside, clearly Shakespeare knew the power of the sun to influence human mood. Without his love, whatever the season, the "leaves look pale, dreading the winter's near" (l. 14).

Shakespeare's soft apocalypse of lovesickness, carried on a solar conceit, feels cozy compared to literal apocalyptic visions of the sun's extinction. Following the Tambora volcano eruption in 1815, which altered the global climate especially in 1816's oddly cold summer, Lord Byron wrote the poem "Darkness." This pocket-sized epic poem imagines the sun's death, the earth's plunge into deep dark and cold, and the violence of human decline. The vicious pathos of the poem, which includes ecological and religious desecration, murder, and cannibalism, is an emotive coating to what is fundamentally a state of global energy starvation. "All hearts / were chill'd into a selfish prayer for light," so the people burn grand palaces and humble huts "for beacons; cities were consumed"; once the cities burn down, "Forests were set on fire—but hour by hour / They fell and faded— and the crackling trunks / Extinguish'd with a crash—and all was black" (19–21). Without the sun's energy, earth dies, joining the blank realm of space matter that seems to make up most of the universe: "The world was void, / The populous and the powerful was a lump, / Seasonless, herbless, treeless, manless, lifeless— /

A lump of death—a chaos of hard clay" (69–72). The word "chaos" might imply primordial possibility, like the unformed matter in Genesis whipped into order, but Byron's chaos is earth's end-game, won by a deified, feminine Darkness: "She was the universe" (82). All the biomass consumed in the poem's hysteria—the forests, the (wood-built) cities—proves to be a brief epilogue following the sun's death. Without the energy of our star, the universe becomes a cluttered vacancy, and the first casualty of its extinction is the human mind, then human society, then everything anarchy can consume.

We may be thankful, then, that Byron's prophecy will probably not occur for another 6.5 billion years—more time than the earth itself has existed. Byron's work is a nadir of sun poems. What about the harnessing of solar energy in literature, not through the usual vector of plants, but through technology? Passive solar has been used since antiquity to heat vats of water, to warm dwellings with dark roofs and strategic window design, to encourage plant growth with dark ground covers. Yet only within the last century have we developed methods for capturing solar energy directly. Solar power in scientific theory coevolved with electricity: in the 1880s, Heinrich Hertz discovered the photoelectric effect and Edward Weston patented the solar cell. But there was plenty of coal, oil was just being discovered, and biomass is as perennial as the grass.

Since its inception, solar innovation has been hobbled by weak demand. Real innovation in sun-capturing technology did not come about until the 1950s, when American and Soviet space programs started using silicon solar cells to power satellites in space. The OPEC oil embargo of the 1970s scared Western consumers into invention, and Exxon Corporation developed a more affordable solar panel that could be applied to sublunary tasks. Since the 1970s, several technological generations have passed, yet the entire industry suffers from anemia due to fuel alternatives. Now, climate change serves as a spur to innovation, and photovoltaic technology is bolstered by strategic energy development on an international scale. Paint and fabric-based solar cells are in development, promising to make solar energy part of the everyday.

The literature of direct solar energy is surprisingly sparse. Perhaps due to its relatively recent advent, its low environmental impact, and its rather dull method of passive generation, solar energy tends to appear only in science fiction and utopian works, and mostly as a symbol of general technological advance. In Jules Verne's 1865 novel *From the Earth to the Moon*, innovators have modified existing artillery to make an enormous cannon that blasts three audacious cosmonauts into outer space. Their aluminum ship is lavish in the Victorian style, with all the comforts of home, including excellent Burgundy and twelve-foot ceilings. It also has skylights and floor windows, through which solar energy warms the vessel to their satisfaction.

Despite the extremes of cold in space balanced with an equal threat of too much solar energy roasting them like potatoes in their aluminum jacket, the men find a cozy medium that echoes the comforts of our Goldilocks planet. The rational Barbicane, who spouts scientific calculations (there is a poet, too, to complement him), contemplates the economic benefits of a free sunlight bath. The ship was launched and runs on fossil fuels, but a little solar boost helps their budget and their mood. This is passive, rather than active, solar generation because the aluminum receives the warmth of the sun's rays but it does not convert the rays, via photovoltaic cells, into another useful form like electricity that might drive the ship. Michel Ardan, the poet, makes Barbicane's economic calculus into the pastoral by germinating peas in the warm window. The ship modulates the gift of solar energy, and the cheerful microcosm proceeds to the moon.

In real outer space, earth satellites, Mars rovers, and probes like the one sent in 2014 by the Rosetta spacecraft to land on the rock-ice comet 67P, all run on active solar energy. Sadly, the probe that successfully soft-landed on the comet bounced into a shady bowl, requiring it to log-off until brighter days, which came in June and July 2015 when the comet drew closer to the sun. The shade-starved probe illustrates the role of luck within severe energy gradients in space. As on earth, shade spells death for organisms and machines that ingest sunlight. Innovators are working with prototypes of spacecraft that might operate on solar power generated from vast arrays of unfurled photovoltaic cells.

One short story by Isaac Asimov, "The Last Question" (1956), gives solar power a starring role in a secular, techno-Genesis. Asimov's dual infatuation with incipient space-age technology and with the new room-sized computers (first developed in the 1950s) melds into an intriguing deep future forecast of our solar system. Multivac is a supercomputer that has evolved through human generations into the main motor of civilization. It calculates and engineers all human endeavors. Multivac is a machine so large in size and complexity that individuals only tinker with the surface: "They fed it data, adjusted questions to its needs and translated the answers that were issued" (170). Eventually, Multivac figures out how to harness solar power, and this technology transforms the world: "The energy of the sun was stored, converted, and utilized directly on a planet-wide scale. All Earth turned off its burning coal, its fissioning uranium, and flipped the switch that connected all of it to a small station, one mile in diameter, circling the Earth at half the distance of the Moon. All Earth ran by invisible beams of sunpower" (171).

Two engineers marvel at the infinite power harnessed by this small solar satellite, but then remember that infinity is a concept, not a reality. Our sun, too, is mortal. When the engineers ask Multivac how to fight entropy to restore the old sun to youth, the computer replies "INSUFFICIENT DATA FOR MEANINGFUL ANSWER" (173). The story travels forward twenty thousand years to an era when the megalith Multivac is now everyone's personal computer, population is exponentially larger, and humans have colonized other planets. Still, the entropy question is unanswered. Take another mind-bending leap forward, and humans have colonized many galaxies, but the sun is dying. Forward ten trillion years, and the universe is dying; for every star created by the universal computer, one thousand aging white dwarfs are snuffed out. Entropy approaches maximum, and humans have melded with their computers to become cyborgs. All data has been collected and all possible combinations have been considered, and finally the computer discovers negentropy—the reversal of thermodynamic decay. The story ends with the neo-Genesis:

> The consciousness of AC encompassed all of what had once been a Universe and brooded over what was now Chaos. Step by step, it must be done.
>
> And AC said, "LET THERE BE LIGHT!"
>
> And there was light. (183)

In less than five thousand words, Asimov takes us from 1950s sparkly technology to the astonishing eschatology of a computer god forming cosmos from chaos. The end is the beginning, God is the machine, and entropy is defeated when chaos is organized into a cradle of energy. Eternity and infinity are conceptual dimensions. Solar literature reminds us that even the author of all life on earth— the sun—is mortal. Science fiction creates outlandish strategies, such as cosmic migration, for surviving this ultimate apocalypse.

As solar PV cells become cheaper to produce, they integrate into civil infrastructure. When solar generation is quotidian, it energizes everyday civic spaces rather than fantastic space machines. Ernest Callenbach's *Ecotopia Emerging* (1981) was early among fictional visions of a solar utopia freed by affordable PV technology from Regan-era fossil fueled business-as-usual. Ian McEwan's *Solar* (2010) starts on a promising premise—that physical science will invent a mechanical technology that achieves photosynthesis. Such dreams would allow us to fix food directly from sunlight, like plants do, but on a global scale that liberates humans from all energy rations. McEwan's antihero Michael Beard, the barnyard swine variety of Nobel laureate, holds sinecure in numerous scientific foundations. A lowly postgraduate blends physics with biomimicry to draft plans for artificial photosynthesis, which, after a series of dastardly acts, Beard appropriates, patents, and privatizes into his own solar empire in the New Mexico desert.

McEwan cannot detail the mechanism that unlocks this holy grail of infinite energy, but even the narrative's attention to that premise is short-lived. Instead, the book continually sinks to Beard's feckless sexual and gustatory appetites, deconstructing the paragon of the elite scientist down to a pack of vying animal urges in a series of repetitive set pieces. Climate change, a famously difficult-to-tell narrative with billions of actors and superhuman time and spatial scales, gets shorter shrift than the curves of female bodies Beard enjoys (Garrard 124). Where the fetish object could have been the elegant flow lines of photosynthetic assembly, McEwan details a series of sexual parts and Beard's irritation on finding them attached to willful women. That makes for a small (but long) story about a fat and maladjusted old geezer whose penis plots the course through the next lavish cocktail party. These antics end as we might expect, with not a single ADP reenergized to ATP by the titular fuel source.

Beard's physical and moral grossness is a social allegory of decadence and decline. His professional behavior models crony capitalism where corporations battle to the death over market share and intellectual property but do little public good in their race for private profit. Stemming the swell of self-interest are government foundations that research science for public benefit, and the novel's denouement throws out the possibility that through reams of litigation the solar–photosynthetic epiphany will eventually benefit earthlings in general. But *Solar* is a character study more interested in weaving an elaborate comeuppance for Professor Beard than in envisioning a world in which physics learns from nature how to save ourselves, and our souls.

Beard's personal behavior invokes our terror of the great maw—the mouth that is never filled, the appetite ever-cycling between over-satiated and, in Beard's term, "pre-hungry." As an allegory of the bestial appetites fossil fuels have infused in industrial humans, *Solar* hits home. But it projects nothing forward. For all its charisma, solar power—ubiquitous and silent—still waits for its transcendent novel.

IV Atom

Splitting the atom is our most audacious way to capture energy. The brutal architecture of enrichment and fission, the dream of infinite carbon-free energy, the threat of the fusion bomb, the leaks and sizzles of radiation, the gnarly conundrum of nuclear waste—all of these tropes of nuclear power seize the popular imagination. Nuclear scientists discovered how to control nuclear energy in the 1930s, following Einstein's proposal that mass could be converted into energy, a full two decades before biologists understood the biochemical pathways of photosynthesis (Smil 2). Though it is a thoroughly modern form of energy, capturing nuclear power is comparatively simpler than converting sunlight into sugars, an alchemy for which we still need plants.

Two pathways to energy—fusion and fission—take opposite approaches to manipulate atoms and their isotopes. Fusion, the nuclear reaction that powers both the sun and the hydrogen bomb, unites different isotopes of hydrogen to release an enormous amount of uncontrolled energy. In eighty years of research, we have not found a way to harness this explosion in a contained space to generate electricity. Fusion can only destroy with the most obscene energy boom, producing scenes of apocalypse first witnessed in 1952 with the testing of the fusion bomb "Ivy Mike" in the Pacific Ocean. With nuclear fission, a large, unstable isotope, usually Uranium 235, is blasted with a neutron that breaks up the atom and releases controllable energy. Fission of uranium powered the nuclear weapons that destroyed Hiroshima and Nagasaki in 1945. Fission creates a lot of radioactive waste, the irradiated by-products of nuclear reactions. In the mid-twentieth century, the world regarded the bomb with terrifying, almost religious, awe; only in the latter half of the century did nuclear fission emerge as a relatively stable way to generate steam that turns turbines for electricity. This latter, electrical, atom has a surprisingly short half-life in literature considering that a large fraction of our electricity comes from nuclear power. The bomb and its fallout dominate the dire literature of nuclear fuel.

The mystique of nuclear power emerged out of a frustration with older forms of energy whose limits and liabilities had become manifest by the twentieth century. The scientific jargon that naturalized nuclear energy starting in the 1950s has had a powerful effect on public perception of its true benefits and dangers (A. Wilson 259). By the middle of the twentieth century, the public was susceptible to a rhetoric of promotion and mystification. Biomass's severely limited energy, dependence on arable land and stable weather, and mountains of fecal by-products had encouraged the industrial conversion to fossil fuels. Coal's dangerous mining, killing pollution, and mechanical working conditions had stained the nineteenth century with the impurities of a high-energy economy. Nuclear fission initially suggested energy that was nearly free, completely clean, fully controllable, and functionally infinite (Smil 255). In a 1953 speech, President

Eisenhower used the slogan "Atoms for Peace" to wash away the specter of the bomb so recent in global memory, and to redress the resultant "nuclearosis" that the US government saw as a public relations obstacle to nuclear technologies with belligerent as well as peacetime applications (A. Wilson 272).

Even before the Manhattan Project and the development of the Hanford Nuclear Site to enrich uranium for the bomb, science fiction writers in the 1930s and 40s used nuclear energy as a metaphor for limitless power, and therefore social and technological revolution, promises that for the most part have been delivered stillborn. In a 1945 book *Atomic Energy in the Coming Era*, David Dietz forecast that cars would soon be driven for a year "on a pellet of atomic energy the size of a vitamin pill" (Nye 201). Though science fiction is a powerful genre for predicting and even laying the groundwork of future technologies, these stories promote extravagant nuclear doohickeys still far beyond our science. Don Stuart (pen name of trained physicist John Campbell) wrote a series of stories in *Astounding* magazine describing atomic power plants that allow humans to zip around the galaxy, figuring the macrocosmic solar system as simply a very large atom that could come under human control once we mastered the art of fusing the microcosmic atom. In his *Foundation* trilogy, Isaac Asimov imagines miniature atomic devices that replace weapons in a utopian future where nuclear technology supplants social and economic reform as the force of improvement (Berger 121–3).

In the 1944 short story "Lobby," Clifford Simak blames bureaucracy and sabotage, not technical hurdles or radioactive waste, for suppressing the ambition of a nuclear power booster named Cobb. His company, Atomic Power, Inc, aims to provide the consumer with "power that will free the world, that will help develop the world. Power so cheap and plentiful and safe to handle that no man is so poor he can't afford to use it" (93). Simak's economic angle is a surprising thesis for science fiction, and it cuts both ways: if nuclear power is cheap and abundant, what will happen to competing power industries that use fossil fuels and hydropower? Cobb's critics fear that deploying atomic power is equivalent to "letting loose economic chaos. You'll absolutely iron out vast companies that employ hundreds of thousands of men and women. You'll create a securities panic, which will have repercussions throughout the world, upsetting trade schedules which just now are beginning to have some influence toward a structure for enduring peace" (94). To bring down his business, journalists extract Cobb's admission that there are unknown effects in nuclear power and the subsequent headlines read that it is inherently "dangerous." A rabble of politicians collude to blow up Cobb's plant, an act of vandalism that kills one hundred workers. But Cobb wins the day: knowing their guilt, he blackmails these politicians to support a redoubling of development funds, and the Pollyanna plot concludes with a vision of a global state run by specially trained political scientists pursuing "balanced scientific government ... Men of science will govern, running the world scientifically in the interest of the stockholders— the little people of the world" (102). This techno-utopian ending is typical of mid-century science fiction, and nuclear stories populate a particularly fanciful subgenre.

Nuclear power becomes a force of democratic development that cuts through human hierarchies based on differential access to resources. The king and wealthy merchants controlled the fleet of wind-powered trade vessels; feudal aristocrat owned the great forests of the medieval world and the animals within them; the industrial baron used corporate law and sheer aggression to gain control over coal and petroleum. In Simak's postmodern nuclear utopia, simple abundance renders access universal and shifts the nature of wealth and work away from capitalism and toward an idealized not-for-profit global collective. Cobb has solved a fundamental challenge to survival—to gather enough energy from the environment—and the success is not meager subsistence but a luxury of supply. The only victims of this system are those whose wealth comes from high-priced energy from other sources, a decidedly unsympathetic clique of John D. Rockefellers. This short story's publication date, 1944, is the last year that the public could be innocent of the dystopian possibilities of atomic energy.

Psychological stress results from the atomic proximity between energy and annihilation. Deploying a quirky technology that has the power to incinerate a large swath of the inhabitable globe carries with it a mental burden also explored in mid-century. Robert Heinlen's 1940 short story "Blowups Happen" takes us into the atomic power plant of the future, which operates in a lockdown state in which technicians are constantly observed by psychologists for signs of cognitive meltdown—their leading occupational hazard. The stress of maintaining the machine in a "self-perpetuating sequence of nuclear splitting, *just under the level of complete explosion*" makes for rapid turnover and rabid suspicion among workers (105). The justification for such a dispiriting work environment is that the high-energy economy needs and demands atomic power.

This futurist society is unsated by its vast "sunpower screens," which "had saved the country from impending famine of oil and coal, but their maximum output of one horsepower per square yard" limited solar power's ability to provide enough local power to industrial centers (114). The story runs scenarios on how the nuclear generators could be exported to space to spare humans in the event of an explosion. This, it appears, is a better solution than auditing the volume of energy consumed by a late industrial high-tech society that cannot change its habits despite the exhaustion of fossil fuels and the privations of solar and other renewables. Three quarters of a century later, we find Heinlen's vision of energy scarcity scarily accurate, particularly when combined with the stressor of climate change linked to burning fossil fuels. Thankfully, we have accomplished gains in efficiency that partially offset an increasingly plugged-in globe, but these gains are nowhere near counterbalance without fossil fuels.

The nuclear war machine existed in fiction before it was so heavily deployed against Japan at the end of the Second World War. Science fiction of the early 1940s predicted many of the secret details of the Manhattan Project and conjectured how the military applications of the atom would lead to the arms race of the Cold War. At times, the fictions brushed so close to the truth that the military suspected secret intelligence leaks. In 1944, Cleve Cartmill published "Deadline" in *Astounding Science-Fiction*; within a few days military intelligence agents knocked at the publisher's door, demanding to know who had leaked information. In the end, Cartmill's imagination and a bit of luck are credited for passages like "They could end the war overnight with controlled U-235 bombs. They could end this cycle of civilization with one or two uncontrolled bombs. And they don't know which they'd have if they made 'em. So far, they haven't worked out any way to control the explosion of U-235" (78).

Ian Fleming's James Bond dynasty came of age in the Cold War era when control of nuclear weapons was a distinction of the world's superpower states, the Soviet Union and the United States. Always the patriot, Bond's adventures bring British military power to the field in a nostalgic nod to his country's historical geopolitical dominance. The Bond series makes a fetish of high-tech weaponry by deploying a character, Q, whose sole purpose is to arm 007 with chichi gadgets. Q does not dabble in nukes, though: bombs are the toys of the bad guys, often rogue actors who steal from the arsenals of the world powers. *Moonraker, The Spy Who Loved Me, For Your Eyes Only, Thunderball, You Only Live Twice*, and *Goldfinger* all maneuver along this basic plot line. 007 uses his personal armory, guile, and sexy intrepidity to outfox more ham-handed villains with fingers on red buttons.

Written for entertainment rather than any serious political or scientific discourse about nuclear technology, Fleming's novels and the succession of movie adaptations often trivialize the effects of radiation. The fiction of nukes-for-fun parallels 1950s British pronuclear weapons propaganda that assured citizens that fallout from radiation was not a serious concern. A 1955 secret government report concluded that fallout from a nuclear attack could devastate the British landscape and bring down the state itself, but the civil defense program publicly maintained that nuclear war was survivable (Laucht 364). As escapist fiction, the Bond series makes a thrilling comedy out of the nuclear terror that haunted global politics throughout the Cold War.

In Ian Fleming's *Moonraker*, Bond diverts the nuclear rocket that the villain Sir Hugo Drax had intended to explode in London; instead it explodes over the English Channel. To his boss M he expresses concern about "Radiation and atomic dust and all that. The famous mushroom cloud. Surely that's going to be a problem" (183). M assures him that the Geiger counters have stayed cold near London: "The cloud's got to come down somewhere, of course, but by happy chance such wind as there is is drifting up north" (183). We never hear how the Scots cope with the fallout breeze.

The most infamous scene of nuclear fantasia comes in the film version of *Dr. No*, which altered Fleming's book to introduce a nuclear-powered laboratory on the villain's tropical island. 007 and Honey Rider (yes) wash radioactive dust from their bodies in Dr. No's "decontamination chamber," and a squeaky-clean Bond compliments the villain by saying "It was a good idea to use atomic power ... I'm glad you can handle it properly." Not so, as the film ends with a core meltdown that kills Dr. No in a vat of reactor coolant, but the disaster is treated as a controlled implosion: no traces of nuclear fallout remain (Laucht 366). The freshly washed Bond and Rider wallow in pleasure on an idle speedboat, having released the tow rope of the rescuing British navy. Bond's potency overshadows even the dread apocalypse of the meltdown.

By the 1980s, after the scare at Three Mile Island and the disaster at Chernobyl, commissioning new nuclear power plants had mostly ceased in the developed world. The apex of the nuclear age coincided with the height of the Cold War, and nuclear fission for electricity suffered by association with its incendiary cousin, fusion. These two nuclear energies brought together opposite and scary aesthetics of apocalypse: the slow violence of radiation exposure and the instant annihilation of the modern bomb (Nixon 49). Literature has made much of these antipodes of disaster.

Corporate rhetoric in the wake of disaster is a flashpoint in contemporary poetry. In "Tar," C. K. Williams complains the morning after the Three Mile Island incident: "We still know less than nothing: the utility company continues making little of the accident, / the slick federal spokesman still have their evasions in some semblance of order. / Surely we suspect we're being lied to" (ll. 8–10). In his poem "Utilities Advertisement in the Wake of Three Mile Island,"

Gary Metras cites an advertisement that obfuscated the threat by comparing it to other sources of body damage: "For radiation consider / x-rays, microwaves, the sun. / This is a matter of risk assessment. / No available energy source / is risk-free" (ll. 12–16). Metras turns this found poem into a jarring clash of messages in which the smooth rhetoric of control is riddled by doubt. Claims of "candor and openness" from corporate and government stakeholders lay the way for a parental request to the people: "This is a time to avoid emotionalism" (ll. 1, 4).

Kathleen Flenniken's *Plume* (2012) is a verse autobiography of her childhood near the Hanford Nuclear Reservation. It develops a leitmotif of redactions in which official statements are blacked down to terse protest messages. She quotes a sentence from the Atomic Energy Act of 1946 that shifted America's nuclear industry from military to civilian control: "The dissemination of scientific and technical information relating to atomic energy should be permitted and encouraged so as to provide that free interchange of ideas and criticisms which is essential to scientific progress" (21). Flenniken picks pith from the prose and blacks out the rest: "Fear / could end / a critical scientific / program." Another redaction takes Robert Oppenheimer's long comments on scientific error, the need for open testing, the critique of nuclear programs, and compresses them into the phrase "Science / vetted / in ignorance / may incite / chaos" (31). These poems distill the noise of all the clashing stakeholders and, with a blanket of ink, set in relief the essence of the problem: cultivated ignorance.

Russian poet Lyubov Sirota examines the word "Radiophobia," a term deployed by Soviet officials to blame hypochondria for the sickness of thousands of citizens exposed to Chernobyl's radiation. The euphemism is only "television bravado" meant to turn the people's fear back onto themselves, to make them question their own sanity in a time of mass hysteria (l. 30). Sirota considers the survivors as "seers" who clearly perceive government cover-ups, and are suspicious of official folderol calculated to distract and obscure the realities of a post-meltdown world. In the end, because of its power to unite the people in collective resistance, "Radiophobia might cure the world / of negligence, greed and glut, / bureaucratism and spiritual void, / so that, through somebody's good intentions, / we do not mutate into non-humankind" (ll. 45–9). Press releases are a tangle of contradictions and withholdings. From the burbling brook of corporate palaver drip a few drops of truth. In poetry, that truth is often a validation of the unknown, a recognition that we are in the middle of an experiment, exposed. Once the atom is split, the world is more insidiously mysterious, less controlled, riskier. We have entered the era of risk management.

These poetic reactions to real-life nuclear disasters are the repudiation of utopian science fiction that would make us omnipotent, with personal fission devices in our pockets. From that egocentric dream we wake to a reality of crushing risk. Our limits are not only restrictions on the energy we can produce and consume, but also the limits of endurance as the organic body is exposed to radiation. June Jordan's poem "Directions for Carrying Explosive Nuclear Wastes through Metropolitan New York" conveys the dread volatility of new elements we have created and that exist at large, not safely buried under Yucca Mountain, but in the middle of our largest cities. She commands the drivers not to pass other cars or drive in the rain, snow, or dark: "Go slow. / Do not brake suddenly or / otherwise. / Think about your mother / and look out for the crazies" (ll. 19–23). That last statement tears apart the control of the earlier lines and leaves the reader questioning who the crazies might be—terrorists, bad drivers, or ourselves, the ones who allow trucks of nuclear waste to accumulate anywhere in the world, let alone our most populous city.

Nuclear accidents expose the body to innumerable threats. Invisible radiation becomes apparent in a chilling range of physical affects, from acute radiation sickness to emergent cancer clusters to the infusion of hot atoms into agricultural goods. In "Tar," C. K. Williams's "Where Were You?" poem written the first morning after the Three Mile Island partial meltdown, a work crew reroofs his building oblivious to the emergency "a hundred miles downwind" (l. 5). Lacking floods and fires and mushroom clouds, thus unable to fix his anxiety on any apparent disaster, Williams focuses on the roofers whose "brutal work" is "harrowingly dangerous" (l. 12). The filth of it transforms the workers "so completely slashed and mucked they seem almost from another realm, like trolls" (l. 19). Recalling the vigil of those days in August 1979, Williams wonders why his clearest memory is of the irrelevant roofers, more than President Carter's visit to the plant. The workmen, "slivered with glitter from the shingles, clinging like starlings beneath the eaves" are burned in his memory because their dirty labor slips into a metaphor for the exposure inherent in nuclear power (l. 32). We cannot perceive radiation—it leaves no mnemonic behind—but we can see spent shingles and crumbling tar in chunks "so black they seemed to suck the light out of air" (l. 33). The neighborhood children use the sidewalks as a canvas for their tar art, "obscenities and hearts," the graffiti of a vulnerable, unknowing populous (l. 34).

Jeffrey Hillard takes the body inside the nuclear facility in his poem "The Message," noting the dissonance between the safety signs posted to remind workers of protocol and the reality of their daily dosages. "*A Careless Employee is a Disaster in Uniform. / ... Today, something is different. / I imagine the tar-black walls swelling, / tiers of sludge frothing over the catwalks, / my lead shield coming apart, dust floating past*" (ll. 16, 20–23). Squelching his nightmarish fears, he substitutes the practical sign "*Shower Thoroughly*" for a fountain-of-youth dream: "*All bodily cells replenished here*" (l. 30). These fantasies of nuclear health disintegrate with the final image of the worker's uniform, a synecdoche for his body, used by the company and discarded in a black barrel at every day's end. The man himself leaves only "footprints to be mopped and forgotten" (l. 39).

Washed-away radiation has healing psychological effects, however impossible it is to purge genetic mutations. Sujata Bhatt's "Wine from Bordeaux" ends with a boy near Chernobyl "begging / to be allowed to take a shower / whenever he came indoors / thinking the water / thinking the water would wash / it all off" (ll. 70–5). That worst of all meltdowns conveys a historical before-and-after diptych, where the joy of a prior life is obsessively compared to what passes for life following meltdown. In Bhatt's imagination there is a man who stockpiles wine from before the meltdown because "He doesn't like / to ingest anything harvested / in Europe after 1985" (ll. 22–4). His paranoia is not a far cry from reality: Bhatt tells the true stories of a man "who got himself sterilized ... because he was convinced / his chromosomes were damaged"; of pregnant women in 1986 who "didn't know what to eat"; of parents who "scrutinize their children / with fear, wishing they could supervise / the health of every cell"; of clusters of miscarriages and children who "suddenly / become sick with leukemia" (ll. 42, 44–5, 50–1, 53–5, 61–2).

The damage to human bodies extends to their agriculture; Mary Jo Salter observes in her poem "Chernobyl" that in Rome "Milk gurgled down the drain. / In Wales, spring lambs were painted / blue, not to be eaten / till the next spring when ... Still tainted, / they'd grown into blue mutton" (ll. 16–20). Mandatory testing of radiation in sheep who graze on contaminated highland in the United Kingdom only ended in 2012, a full twenty-six years after the meltdown, but media attention waned long before then ("Post-Chernobyl"). Salter's point is that the disaster only consciously oppressed Europe so long as it was the subject of media hysterics; once "the story at last was dead— / all traces of it [were] swept / under the earth's green bed" (ll. 26–8). Detailing agricultural waste from that disturbed era brings an ecological perspective to the economic account of losses, unifying the toxic load of all bodies, human, nonhuman, and planetary.

Contamination does not always spell death, though: it can be an avenue for surprising turns of evolution. Reviewing the effects of nuclear radiation in the long term, Alan Weisman surveyed the neo-wilderness that has evolved in the environs of Chernobyl for the last quarter century. The thirty-kilometer radius Zone of Alienation is now thickly forested and peopled with bear, birds, deer, boars, lynx, moose, and wolves (215). Radioactive habitat is terrain these wild animals could not have occupied before, and Weisman cites radio-ecologists who believe that "typical human activity is more devastating to biodiversity and abundance of local flora and fauna than the worst nuclear power plant disaster" (217). Though the lifespan of these animals has proved to be shorter than their forebears, some populations show tolerance to radiation thanks to favorable mutations. The hot habitat—even one heated by radiation—is a cradle for creative evolution. Seen from this perspective, meltdown is an environmental punctuation similar in kind to natural disasters, but even more potent in its ability to create genetic mutation, the agent of evolution.

In *Four Fields* (2015), Tim Dee scans the Chernobyl exclusion zone with biologists who have studied the epidemiology of fallout ever since the 1986 meltdown. Dee enjoys the huge population of swallows that swarm over the pastoral landscape: "Their lovely chittering flights, warm, sociable and unhindered, seemed to come from another time and place, the best of a summer farm thrown up into the autumn air above a forest" (182). But their search for aberrations quickly yields phenotypic evidence of mass mutation: tumors, partial albinism, asymmetric feathers, malformed beaks, toes that point the wrong way (185). The villages abandoned for a quarter century sink into the land, their pointy bits snapped off, and the swallows "shake from these places like dead leaves or brick dust.... By their flights and calls through this air the swallows advertised it as habitable. But the swallows of Chernobyl are sick. The air might be their zone but the insects they take from that air pull the birds heavily down to earth" (182). Stuck in a sick food web, the birds' habitat transmits hundred times the background radiation of nuclear-free zones, steeped in the hot Plutonium-239 that has a half-life of 24,100 years (189). The data shows only 28 percent of Chernobyl swallows return to their birth brood to reproduce, versus 49 percent of the normal population (186). Birds that do not return to brood are probably dead.

Dee visits the remains of the Chernobyl worker's village Prypiat, built in 1970 as "a space-age city for an atom happy planet," but civically alive for only sixteen years (207). "Down the wooded streets of Prypiat's arrested past you are bowled into the aftermath of man, into a future that has already arrived. I have been nowhere else that has felt as dead as here, been nowhere that made me feel as posthumous. And the strangest thing is that in this house of the dead, the dead have gone missing" (209). The visual cues are as poignant as any contemplation of ruin, from Mayan temples eaten by the jungle to Ozymandias commanding an oblivious desert. But the feel is visceral and uncanny, not metaphysical: "I began to imagine I could feel my own porosity, the soft membranes of my body, and the weather moving through me, the air coming inside and laying ash on my tongue" (190).

This presence of material hazard cries out as absence in the emotive landscape of the fallout zone. Dee looks to the sky to escape the wasting wasteland and imagines more of the same: "The moon was alone in the sky when we finished. Half of it, only" (201). A quarter century after Chernobyl, the exclusion zone has reached an illuminating stage of decay, where human remnants are strangely intact but shifted in context from elements of life to those of some incalculable, sprawling afterlife. We attempt to take stock of the legacy through numbers of the dead and poisoned, ratios of mutation, Geiger counters, and radiological half-lives of more than twenty-four thousand years (humans were still ten thousand years preagricultural if we reach that far into the past). But the most moving aspect of the nuclear fallout experience is not in the numbers but in the pathos: the eloquent remnants, the sixth sense of body invasion that sojourners may try to write off as paranoia.

Postapocalyptic fictions like Lewis Padgett's short story "The Piper's Son," and more cryptically the novels *Riddey Walker* by Russell Hoban and *The Road* by Cormac McCarthy, also imagine what remains in the doldrums of nuclear winter. This wildly popular genre animates all of our latent fears about the nuclear chaos that will bring down our high-energy, economically stratified society. The two novels work on the level of mythology and postindustrial subsistence demanded by a highly hostile new environment; the rebel atom is a catalyst but not much of a character. But in Padgett's short story, we find a genetic fantasy intimately related to radioactive exposure. The bomb has destroyed Denver and Chicago, and those who lived in the environs were subject to a genetic mutation that gave them powers of telepathy and incidentally made them hairless. The Baldies' superpowers make them pariahs to normal humans, and they have split into two groups: those who suppress their telepathy, wear wigs, and genuflect to the ruling orders; and those who embrace their capabilities.

For public safety, members of this latter group have been declared insane and committed to asylums. A father Baldie wants to save his son from this fate after he sees the boy growing increasingly arrogant and antisocial. The boy fantasizes that he is a "Green Man, a figure of marvelous muscular development, handsome as a god, and hairless from head to foot, glistening pale green" who defeats the jealous, impotent "hairy gnomes" (45). A tender but tense relationship develops between father and son that would, in standard fiction, revolve around the son's sexuality or social anxiety. But this is science fiction, and telepathy and the Green Man emerge as the superpowers of a super energy working at the atomic level on our genome. Padgett's quiet, thoughtful story about a son's confusion and a father's counsel has evolved into a set of comic book characters like D. C. Comics' Green Man superhero, Marvel's heroic Incredible Hulk, and villainous Radioactive Man. Bart Simpson's same-named hero has eyes that beam nuclear energy. With its nuclear power plant ill-secured by Homer Simpson, Springfield is an ideal locale for radioactive citizens with delusions of grandeur. These fantasies of radioactive exposure as empowering rather than degrading show an enduring cache of wild comedic optimism for nuclear technology that is the bipolar response to the laments and jeremiads of fallout literature.

Also reflecting on toxic body load, but with opposite mind-body affect, Kathleen Flenniken recalls her childhood near Hanford Nuclear Site in Washington State, the main facility for producing weapons-grade plutonium between 1944 and 1987. As an elementary student, she was required to "do a little for our country" by having her body tested for radiation in a portable body counter (l. 5). Imagining America as a soothing pastoral of parks and lawns, she enters the machine "like a spaceship and I moved / slow as the sun through the chamber's / smooth steel sky" and whispers her warped patriotic creed infused with a pun, "so proud to be / a girl America could count on" (l. 19–21, 24).

The innocence of "Whole Body Counter" evolves into experience with the next poems in the sequence, "Plume" and "To Carolyn's Father." The first is a block poem where the words threaten to escape the containment vessel of the column. "Plume" imagines the body of the landscape infected by "poison / yes it is moving / to the river yes / it migrates / between grains / down to / saturated sediment" (ll. 40–6). Hanford is situated along fifty-one miles of the banks of the Columbia River in Washington State, the largest river in the Pacific Northwest. It meanders westward from the scrublands of the east toward Portland and the Pacific Ocean, and Flenniken imagines the unspeakable "it" on a course "trailing its / delicate / paisley scarf / … it will move / downstream / to the river / yes the river / will take it in" (ll. 58–60, 67–71). The highly abstract toxicity has a beautiful molecular structure imagined as paisley, and its movement in lady's scarf culminates in a sensual fusion of poison and fluidity.

The beauty of this union is all the more unsettling for its pernicious turns. Hanford is the most polluted nuclear site in the country, officially stated (in passive voice) to be contaminated with "hundreds of billions of gallons of liquid waste … generated during the plutonium production days. These liquid wastes were disposed of by pouring them onto the ground or into trenches or holding ponds. Unintentional spills of liquids also took place" ("Hanford Cleanup"). The intimacy of Flenniken's poetry evokes the incredibly complex molecular mixing between waste, soil, and water. The cleanup process itself is somewhat simpler, though gargantuan in scale: the Environmental Protection Agency is supervising a landfill operation in which mountains of solid waste will be entombed in cement cells and liquid waste will be fused with liquid glass, cooled to solid, and buried in sealed cylinders ("Hanford Cleanup"). Nature continually stirs the elements, and our legacy at Hanford requires us to confound the natural mingle with postnuclear sarcophagi.

In the next chillingly intimate poem, Flenniken recalls her effort to save Hanford from closure in the early 1970s with a letter-writing campaign. Almost a sonnet, "To Carolyn's Father" begins with the girl called to the principal's office for her activism on behalf of Hanford, which was threatened with closure by President Nixon, "who obviously didn't care about your job" (l. 9). The volta shifts to the infected father's bone marrow cells: "A few gone wrong stunned by exposure to radiation / as you milled uranium into slugs ... / and by that morning Mr. Deen / the poisoning of your blood had already begun" (ll. 12–13, 15–16). Completely unpunctuated, the poem is a run-on of recollection and regret, innocence and realization, which continues the earlier poem's intermingling of rivers and radioactive waste. The former is earthly, the latter is bodily. The radiation is ubiquitous; some of its locations can be pinpointed (in Mr. Deen's marrow), but as soon as it is marked, it diffuses to everywhere in the material and emotional world.

Flenniken's sixty-two pages of postnuclear poems in *Plume* travel through a lifetime of memory capturing the nuclear trope in myriad poses: graceful, lewd, shrouded, and lurid. The final poem, "If You Can Read This," buries each word in a tomb of inscrutable implication. The final four words capture the geographical arc of grief: "[planet (or atom?)] / [traveler] [death] [turn back]" (ll. 8–9). Beneath the emotional grief of these parenthetical cries, she raises the practical question of how we warn future earthlings of the enduring pollution our society has deemed necessary to create energy. Hanford is a void, a lost site that has swallowed the lives of its past workers and turns its destructive appetite on the future. Flenniken never mentions the violent ends of Hanford's plutonium enrichment, the bomb that erased Nagasaki, or Cold War stockpiling. In Hanford, local grief echoes as an implied global lament.

Allen Ginsberg's 1978 chant, "Plutonian Ode," takes its form and grandeur from Whitman. It is a loud lyrical exposé, a radioactive Song of Myself.

Ginsberg's outrage and zeal are inspired by the scientific accomplishment of creating "a new thing under the sun," a repudiation of biblical teachings that is itself a plague of biblical scale: "One microgram inspired to one lung, ten pounds of heavy metal dust adrift slow motion over gray Alps / the breadth of the planet, how long before your radiance speeds blight and death to sentient beings?" (ll. 1, 26–7). The ode declaims against the silence and darkness of nuclear waste caches: "I enter with spirit out loud into your fuel rod drums underground on soundless thrones and beds of lead / O density! This weightless anthem trumpets transcendent through hidden chambers and breaks through iron doors into the Infernal Room!" (ll. 40–1). Taking inspiration from the chthonic Greek myths, Ginsberg identifies a new underworld of eternal warring among the elements energized by the "Radioactive Nemesis" (l. 13). He exposes the covert operations of nuclear munitions plants in the "silent mills at Hanford, Savannah River, Rocky Flats, Pantex, Burlington, Albuquerque" and their support of "Satanic industries projected sudden with Five Hundred Billion Dollar Strength" (ll. 20, 48). The protest poem ends with a call to antinuclear action: Whitman, Congress, and the American people are to "Take this wheel of syllables in hand, these vowels and consonants to breath's end / ... Magnetize this howl with heartless compassion, destroy this mountain of Plutonium with ordinary mind and body speech, / thus empower this Mind-guard spirit gone out, gone out, gone beyond, gone beyond me, Wake space, so Ah!" (ll. 60, 64–5). Ginsberg's barbaric yawp is a syntaxis of helpless fury.

On March 11, 2011, a large earthquake caused a tsunami that rolled onto the coast of Japan, engulfing the Fukushima Daiichi Nuclear Power Plant. The initial loss of power led to a catastrophic meltdown of three of the plant's six reactors. In the days that followed, the failed cooling systems caused the reactors to heat up, and several hydrogen explosions, as well as controlled venting of air and water, released radioactive isotopes. Officials created a twelve-mile exclusion zone surrounded by another ring of concern in which citizens were advised to stay inside. Hundreds of thousands evacuated to nuclear refugee camps. Radioactive water in the Pacific Ocean reached the west coast of North America within a week, and airborne radioactive isotopes from the disaster were detected globally within a month ("Fukushima-related measurements by the CTBTO").

The total radiation released is unknown, but is believed to be a significant fraction of the apocalyptic event at Chernobyl, and the radiation was still not contained in measurements taken in February 2017. The containment and cleanup operation continues to be riddled with emergencies and ersatz protective structures like the shield of permafrost constructed in 2014–5 using hundreds of coolant-flowing pipes driven deep into the ground. The horror and grief of the disaster—the acute effects of the earthquake and tsunami that killed nearly sixteen thousand people, and the radiation that has cast a malaise over the exclusion zone—are further charged by the anguish of a Japanese people uniquely wounded by atomic energy. Hauntings of Hiroshima and Nagasaki ripple through the poetry that has emerged in response to the disaster at Fukushima. Japanese poetry is beyond the scope of this study, but the volume *Reverberations from Fukushima: 50 Japanese Poets Speak Out* (2014) collects the humanist reactions to the most recent crisis. Why, the poets ask, would we take such risks for our electricity?

Today's answer to the question "Why nuclear?" is climate change. This pronuclear environmental stance is an about-face from the antinuclear activism of the 1960s and 1970s that consistently linked energy generation with the bomb. Many respected climate scientists today look to nuclear power as a major carbon-free solution to critical energy demands in the twenty-first century. The first nuclear power plant commissioned solely for energy production was opened in 1957 in Shippingport, Pennsylvania, a locale chosen partly because coal pollution was such an acute problem in that region. Globally, coal is still the dominant fuel source for electricity and it is also still the fastest growing source because of its intense use in developing nations. Coal particulates kill about three million people annually—easily the most deadly fuel for electricity—whereas nuclear energy is the second safest source behind wind power, even after taking into account the high-profile accidents at Three Mile Island, Chernobyl, and Fukushima. (Manufacturing solar cells involves toxic materials that sicken factory workers and their environs.)

France, a nation that gets 75 percent of its power from nuclear sources, has high standards of living and industry but a per capita annual carbon emissions rate of only five tons per year—half the emissions of neighboring Germany, which gets about 30 percent from nuclear sources but relies more on natural gas and coal (*Pandora's Promise*). This discrepancy is all the more surprising since Germany is a world leader in renewable energy from sun and wind. Nuclear power provides constant energy to the grid, unlike sporadic primary sources from sun, wind, and water. Many pronuclear energy policy experts contend that renewables alone cannot replace the global use of fossil fuels without nuclear technology in the portfolio. Since increased access to electricity directly correlates with higher standards of living, our energy strategies are intimately tied with the human right to have access to healthcare, education, and refrigeration. We need to balance the energy needs of billions of consumers with the opposing need to reduce carbon emissions to curtail virulent climate change. As global energy demands increase into the twenty-first century, greater efficiency and wise use will be essential behaviors for all electricity consumers. The shining virtue of nuclear power, its lack of a carbon footprint, is seen by many as too good to pass up.

The optimism about a nuclear renaissance picking up the slack from other sources has waned somewhat since the Fukushima accident. Japan previously gathered 30 percent of its energy from the atom, but afterward all fifty of the country's reactors were temporarily shut down and imported fossil fuels are a stop-gap ("Japan Shuts Down …"). Though doubtless these expensive investments in Japan's infrastructure will be brought back online, national public opinion has pivoted against nuclear power to favor fossil fuels, despite their climate and health impacts. As a global solution to the climate crisis, nuclear energy has its downside. Its plants are expensive to build and complex to license, construct, and regulate. There are continual concerns about waste storage and nuclear proliferation.

On the other hand, nuclear power has the potential to diminish the nuclear weapons threat as Cold War stockpiles are reprocessed to fuel electricity generation. Half of nuclear power in the United States comes from reprocessed materials from nuclear warheads (*Pandora's Promise*). The scruples regarding nuclear power could be fatal to carbon emissions reductions. It takes at least a decade from initial licensing to the time when nuclear power pours onto the electrical grid, and even longer in countries without existing nuclear infrastructure. America's aging fleet of plants is set to retire by 2030, so there will actually most likely be a reduction of nuclear power to the domestic grid, not the kind of paradigm-shifting leap that would reduce the burden on the climate (Ling). Since 1993, the share of nuclear energy in the world has actually decreased (Massey).

None of this is meant to imply that the checkered nuclear era is over. New plants have been commissioned in over a dozen countries clustered in Asia, but also in Russia and the United States. The United States carefully maintains its middle-aged plants to keep nuclear capacity steady as renewables gain market share and coal use declines. The tremendous potential of an atom-powered globe still tickles the imagination of nuclear optimists like it did for science fiction writers nearly a century ago. But the fundamental fears and flaws of nuclear power, and its cousin the atom bomb, permeate literary and environmental responses to this dangerous fuel.

Literature is clearly partisan against nuclear technology, and the chasm separating Pollyanna and Pandora seems unbridgeable. The aura of nuclear oppression is unlike that of any other fuel. Hay and wood are the fuels of seasonal, slow, preindustrial time; coal is sooty, sulky, industrial; wind and water are elemental, tangible, ecological. But nuclear fuel is nothing like any of these fuels. It represents the height of scientific ingenuity and the nadir of apocalyptic horror. Its elements are imperceptible by our senses, making the energetic atom and the violence of the hungry radionuclide ghostly half-presences in our lives. The proleptic drive that pushes us to invent narratives about the future, whether fictional or scientific, is starkly bipolar on the nuclear issue. It is easy to be neutral about hay. But nuclear energy is heaven and hell, contained in the tiny universe of the atom.

9

Conclusion

Physics equates energy with work, a mechanical calculus. The thermodynamics of energy is only the first tier of understanding how modern ecology relies on a series of tricky conversions between fuel and its outlets: work, heat, light, and food (fuel reborn) (Johnson 176). Consider a basic energy-based distinction between two farmers. The mindset of the plowman and his oxen, fueled by biomass, are quite different from the mindset of the farmer atop a massive diesel tractor, even though the final aim, a plowed field, is the same. In relative quiet, the plowman directs and cooperates with his animals, perceiving their moods and health, and observing minutely the turned soil that passes beneath his feet. The soil has tangible, aural, olfactory, gustatory, and visible qualities that he recognizes. His work behind the animals has its own dungy redolence. He may sing a work song to keep the cadence regular and the mood high. His attention is fully dedicated to the unmediated, unsheltered conditions of work, so the vernal shower and the spanning sun inform his thoughts. Growing fatigue does, as well, and perhaps concerns about bad harvests spurred by the memory of them. Patience with himself, the oxen, the plow, and the soil are required, and the talent of patience may assist his higher meditations over ten hours of labor. The scale of his work is confined by the endurance of muscles; a furlong is the standard distance that an ox could plow without resting—about one-eighth of a mile—before turning back to plow the next furrow. An acre is the area one man and one ox could plow in one day—one furlong by one chain (twenty-two yards). Area is defined by energy output/time: the acre converts biomass energy into geography.

As a rough comparison of energy and effort, modern tractors can exceed 350 horsepower and plow a favorable acre in five minutes—about 1/120th of the time it took our traditional plowman. The farmer driving a tractor, citizen of the industrial era, may share some thoughts on soil fertility or physical fatigue with the plowman, but kinship between them is limited. Wendell Berry has argued that conventional farming is closer to mining (the wealth of the soil) than it is to traditional muscle-driven organic agriculture (Nye 193). The machine driver dwells in a space of mechanical noise and fumes, vibrations and jolts, and perhaps even air conditioning and satellite radio. He sits two yards above the soil and rarely touches it. He thinks of space and limits of work in the terms of the fuel tank, and he probably thinks of fertility in terms of petrochemical additives, also deployed by a rumbling machine. He does the work of dozens of plowmen, freeing the (greatly increased) population to work not as farmers cultivating biomass, but as operatives in the numberless tasks of industrial production and consumption that keep us busy in an economy that requires us to work for food, but not to work to produce food.

The modern industrial farmer is more a mechanic and soil chemist dealing in fossil fuels than he is a husbandman and cultivator orchestrating biomass. It's not at all surprising that humans have embraced this much faster method that fobs off hard work onto machines and fossil fuels, nor that we interpret this shift as a kind of natural progressive evolution of human culture. The poignancy comes with what we've lost for the gains, losses that receive very little attention in a mainstream culture designed to value monetary profit under the economic mantra of limitless growth—a kind of anti-ecology.

A suburbanite can observe this distinction between muscular and fossil fuel energy by clearing leaves from her lawn. If she chooses the old-fashioned rake, the work will involve steady physical work, the rustle of leaves and twitter of animals, and the smell of damp leaves in gentle decay. It may take up her entire afternoon, and she may have a sore back afterward. If she chooses the modern leaf blower, the work will be less physical, quicker, deafening, dusty, and fumy. The former involves immersion in the task at hand and perhaps considerable sensory pleasure along with slight discomfort; the latter implies impatience, sensory disharmony, and hastens to finish because the task is noisome.

The end is the same, but the means could not be more different. We usually choose the leaf blower because we wish to avoid physical work, which has come to appear laborious—unworthy of our time when there are so many other attractions. The blower is more "efficient"—in the ideology of time as money— though by calorie count it is much less so. The same contrast of biomass and fossil-driven work can be found inside the home with the broom versus the vacuum, the mortar and pestle versus the electric grinder, the knife and board versus the food processor. These choices—we make them constantly, whether to walk or drive, whether to grow a garden or take a summer trip to Europe, whether to patronize the farmer's market or the Megamart—are all ethical,

ideological, and ecological choices. They add up to our existence, how we chose
to dwell on earth. They may all be reduced down to fuel, or built up to our
experience of what life is. There is nothing inevitable or predetermined in the
ways cultures select and deploy energy; there are only options and choices,
which in retrospect are sometimes confused as destinies. History is filled with
charismatic personalities who appeal to the masses with their optimism, and
the history of fuel shows how industrial optimists from the railway's George
Stevenson to the automobile's Henry Ford touted the advantages of fossil fuel
and its reliant technologies without accounting for its problems. Problems are
more difficult to forecast than profits.

The preindustrial world was constantly pressed against the limits of biomass
and worked to conserve its energy to maintain sustainability. Feasts and
bacchanalias represented the seasonal joy of achieving a margin over the energy
line, permitting the effusion to feast and imbibe as an exception to low-energy
life as usual. (A major exception to the stricture of conservation came with the
first waves of settlement on formerly wild lands, especially in North America
where vast forests were cleared and burned to make way for farms and pasture.)
The industrial world, on the other hand, is designed to consume energy, not just
to sustain life, but in any way that is in the short term economically profitable
or just dumb fun. Fossil fuels are wonderfully indulgent of our evolutionary
instincts to secure calories, modify climate to comfort, and powerfully and
rapidly move our bodies through space. They give us the illusion of being gods.
This illusion is so potent that we struggle to conceive of a world without them,
of what we would be without them. Coal and oil are now as much a part of
our nature, our ontology, as the sowing and the harvest; many would argue that
fossil fuels have fully eclipsed the essential rituals that permitted human survival
for thousands of preindustrial years (botanical, animal, and seasonal knowledge,
access to clean water, physical self-defense). The Industrial Revolution imposed
the most rapid and comprehensive social evolution in the history of our species,
alienating our understanding of the lives of our ancestors just three or four
generations ago.

Solitary, poor, nasty, brutish, and short. There were many essential downsides
to low-energy biomass life, whether of the itinerant hunter-gatherer or settled
agrarian kind. Arable land was strictly limited, conditions were variable, and
populations continually exerted upward pressure against the bubble. Average
bodily comfort in terms of safe food, drink, and pleasant temperature was much
lower. Disease and famine lurked continually. Work was usually physically
demanding, repetitive, and exposed. It was more often dangerous. Margins

were closer because the energy line fell lower on the graph, just skirting above the survival line. Self-generated hay, wood, meat, and grain don't lend much breathing room. In a biomass world in an average year, the buffer from going hungry in late winter was usually thin: several thousand calories of leftover grain, potatoes and preserves in the pantry, lumps of hay converted to meat and eggs. The perennial struggle to survive on Greenland culminated each late winter when the Norse would "pray that the stores of dairy products and dried meat for human food, the hay for animal fodder, and the fuel for heating and cooking didn't run out before the winter's end" (Diamond 231). Jared Diamond's analysis of cultures that fail to subsist has a leitmotif of biomass collapse, lost forests and topsoils especially, which catapulted marginal societies off the land in places as diverse (and isolated) as Easter Island, Ancient Maya, and Norse Greenland. Even in times of peace, among societies well-established in farming and trade, life on biomass involved relentless work, careful provisioning, and inevitable privation.

Now, in any season, the average fossil-fueled individual has billions of calories at her disposal from numerous competing retail outlets, if she has money to exchange. Fossil fuels buoy our high-flying energy line well above survival line, so far above survival that most people in developed nations have never experienced anything life-threatening, except the situations fossil fuels put us in, like a car at highway speed. Where hunger exists, it can be accounted to failures of social protection rather than a lack of available calories. Hardly a vanishing fraction of present-day humans in developed nations would choose to switch their lives with those of ancestors in preindustrial times—the vast majority would remain in the world of Dunkin Donuts and immunization. The ontological losses and the environmental impacts of a fossil-fuel world, issues of philosophy and ecology, measure poorly against its evident comforts enjoyed along every inch of the body. Materialistic ennui cannot compete with deep physical hunger on the scale of suffering. So, as literature tends to idealize the old ways, it is important to remember the variety of harrowing implications of living without gassed-up engines.

Fossil-fuel-driven capitalism satisfies our inherent desires to make ourselves physically comfortable and well-fed and to compete in life's races, large and small, and feel that we have won. From the green light drag race against the stranger next to you, a battle that rationalizes driving a car that gets eighteen miles per gallon, to the lofty 401(k) balance at retirement, we define ourselves materialistically. This label of success reflects number (net worth) as much as number of possessions, and the president and the television and the neighbor and each of us to ourselves ascribes self-value to material success. There's a mountain of self-worth on sale at the Big Box. That's the upside of fossil fuelology. The downside is that in order to afford our ticket, we must work within the cash nexus, usually several degrees of separation from the agriculture that sustains our life. In the organic economy, all work was a first- or second-degree relation to food generating. Work in the modern world has become sedentary, bureaucratic, anonymous, and often trivial in outcome. It dissolves self-identity into corporate conformity. It leaves our human nature wanting physical work, even if our specific bodies have forgotten what that is, or interprets work as a half hour on a coal-plugged treadmill. We are replaceable, expendable, and the result of our work is simply to keep the machine in motion: the factory line, the checkout line, the customer service hotline, the drive-thru lane, the paper trail, the tenure track.

In a biomass world, work was intimately tied with survival. It is one step between an old-time farmer's actions and his supper plate; two steps between for the shopkeeper in town. Work was easy to explain: "I make barrels." Rather than a compulsory exchange of money, goods and skills were traded readily in a barter economy, giving each skilled member of a community a sense of belonging and purpose. At its best, artisanal work in small communities required personal relationships of trust within collaboration, and working interests often translated into civic ones, with prominent skilled craftsmen often doubling as politicians, negotiators, and religious leaders (Nye 54). The limits of energy forced communities into synergetic cohesion and the sharing of wealth, as well as privation. At its best, the preindustrial model is at least as good for quality of life, and perhaps better than the welfare state of a modern social democracy.

In the high-energy information age, job descriptions come in dense paragraphs of special language, and their relation to the basic purposes of life to feed yourself, to feel love and safety, and probably to procreate exists solely in the paycheck that itself no longer has material form. Numbers on a website on a computer screen turn into bread and wine. What alchemy! Your work has nothing to do with producing food or building shelter or preparing for winter. Your paycheck delegates all that to outside channels: dairy farmer, contractor, HVAC guy. So, fossil fuelology has at its heart an irony: it gives us everything evolution has taught us to desire while depriving us of the nature-navigating work we spent long, formative millennia doing in order to survive. We get the stuff without making it. We survive without owning ourselves. Kraft and Cadillac own us, fill us, move us. Disengaging from fossil-fuel culture involves retraining the body to operate without the superhuman infrastructures those fuels have provided, from the excessive speed of the car to the superfluous cool of the air conditioner. This "decoupling" of body from memory is a central challenge of adaptation into the twenty-first century (LeMenager *Living Oil* 104).

Generally, we have acquiesced to the costs to mind, body, society, and nature that fossil fuels have exacted in exchange for the luxuriance of energy. Retrospect often recalibrates the cost-benefit analysis in a more balanced way than contemporary perspectives possibly can. Retrospect often gives more credit to costs, liberated from present concerns of rapid development and continual growth in economies. The steam engine's benefits helped dispel the manifold problems with pollution, disastrous accidents, and social oppression. Another beneficiary of present-tense optimism was the automobile, which initially transformed the American landscape toward access and convenience, but ironically has left a legacy of inaccessibility and inconvenience that we mean to redress with walkable community retrofitting. Nuclear energy optimism has rationalized its costs, too, of pollution and contamination, and perhaps the steepest disaster-to-benefit ratio energy technology has ever seen. A steam engine can blow up and kill a hundred workers, but a meltdown seeps beyond the serious events at Chernobyl and Fukushima to saturate an apocalyptic imaginary on a global scale.

Any "renewable" must be given the necessary time and space to renew itself, and many biomass sources—wood, whale oil, and peat—are fuel sources that have been harvested unsustainably. The exhaustion of one natural resource inevitably excites economic development of the next available energy platform: fossil fuels. Fossil fuels are more densely energetic than biomass, so the great industrial conversion of the last three hundred years represents a "step-up" platform, especially when fossil fuels were relatively easy to mine and drill. Now that is no longer the case. Our next energy platform will likely be some combination of new renewable, new biomass, and nuclear: a portfolio of dispersed, sporadic, and occasionally dangerous sources. Besides, they are not likely to provide as much energy as that provided by coal, natural gas, and oil. From an energy perspective, any step away from fossil fuels is a step down.

Consuming fossil fuels is the very definition of unsustainable. For the last two hundred years, straight through the present and for some unknowable extent into the future, we are defining a radical blip of high-energy frenzy that is the exception in human history. It cannot endure in its present form much more than a few generations, considering limits set upon supply and the climate change that may shake industrial consumerism at its foundations. So, to some extent, past is prologue: renewable biomass will return to the human scene, complementing the contributions of high-tech new renewables from the sun, winds, and tides. This inevitable change, both paradigm shift and paradigm slip, will probably not transform us to itinerant bands of hunters or peasant laborers working scraps of land. Apocalyptic fiction plays with such notions, but the future truth is likely less sensational. Imre Szeman comments: "What we desperately need in our literary present [is] ... narratives that shake us out of our faith in surplus ... not by indulging in the pleasures of end times but by tracing the brutal consequences of a future of slow decline, of less energy for most and no energy for some—a future that might well have less literature and so fewer resources for managing the consequences of our current fictions" (326).

Communities need redesigning. Neighbors need introduction, and individual possessions open to community networks of sharing. Food needs more local sources of production, and higher rates of participation. The world will shrink back toward the scale of muscles, except the internet exchanges that may continue to unify the globe. To a limited extent, the internet allows us to have it both ways: global information, local production and consumption. The best way to avoid tumbling off a cliff is to scale down one careful step at a time. The optimistic tail of the Anthropocene arc envisions human culture proactively changing from fossil sources to renewable ones, and adapting to climate change through active innovation in agriculture and stewardship.

The energy humanities show how labor translates most obviously into energy, but more subtly into myriad movements of history, cultural and ethnic identity, and ecology. For example, music theories suggest that modern jazz music evolved to help attune urban populations with the tempo of electricity, rapid transport networks, and bustling production; the nervous, aimless, improvised jazz brought exosomatic conditions into the body's rhythms (Johnson 73). Literature has enough complexity and depth to enlighten readers on many levels at once—from emotion to history to genre to energy—without collapsing into simplistic admonition and prescription. In this book, the background energy has been brought to foreground, but the characters and their dramas endure, enriched further by the context of their fuel regime. Literature records the historical continuum from annually renewable fuel toward nonrenewable fuel in our recent history, and it preserves the lives and landscapes that have evolved along this historical arc.

William Blake proclaimed that "energy is eternal delight." Seeking to recover the body's pleasures from a censorious Anglican church, Blake's marriage between heaven and hell embraces the diabolical energies of sex, revolution, and individual faith. These higher-order functions result from the simple, intimate act of eating to sustain the body, which is our interface between the soul and the mysterious material/spiritual world that Blake and the Romantics explored. Energy is fungible, transforming from food to agency in the world, and Blake saw that energetic devils were better vessels to embody his ethos of revolt than were prurient angels preening over dead texts. The devils feel both despair and delight, joy and its partner, melancholy. Amid the Industrial Revolution, Walter Pater asked his readers an ontological question: "How shall we pass most swiftly from point to point, and be present always at the focus where the greatest number of vital forces unite in their purest energy?" (197). His answer comes in the very next sentence: "To burn always with this hard, gemlike flame, to maintain this ecstasy, is success in life. In a sense it might even be said that our failure is to form habits: for, after all, habit is relative to a stereotyped world, and meantime it is only the roughness of the eye that makes any two persons, things, situations, seem alike" (197). We might toss this aside as a young man's philosophy, or a young society's, impossible to maintain through the drudgery and repetition of our thirty thousand days on the planet. But in the end it is a matter of spirit and attitude, not age.

At both individual and collective levels, we can cultivate a hard gemlike flame of infatuation with the variety of life if we train our minds to it. Better yet, no high-energy fuels are needed to keep the flame lit. In fact, these fuels, by doing work in the world that we have evolved to do ourselves, often tend to dull the edges, blur the distinctions, and abridge the richness of experience available to the susceptible person. We have worn into a deep groove of high-energy self-absorbed ideology that seems deterministic, but the elements of another, subtler, more ecstatic world have always roiled in the background. By sidelining the fossil-driven ego, we may pour over the beauty of the forms around us, which capture today's energy within the coil of life. The aesthetic of life, made of photosynthesis, hydraulic cycles, and trade winds, beckons stronger than the cadaver soup of the primordial fossil. It is a hazard, to humans and the planet, to lack the imagination that envisions the many ways in which the future might be cultivated. Our lifetimes in this twenty-first-century Anthropocene provide the opportunity to craft a future that brings the best out of fuel.

And for those who love the machismo of a Cadillac shredding the American West, there is a promising future in extreme speed. The Tesla Hyperloop has tested at more than two hundred miles per hour; Maglev trains in Japan have carried passengers at more than three hundred ("Japan"). The twenty-seven-mile test track near Fuefuki harkens back to the Stockton and Darlington line in 1830s England, where Fanny Kemble was wonderstruck at thirty-five miles per hour under a trail of smoke and steam. But tomorrow's hyper-speed normal will use linear induction motors, which can run on the sun.

Bibliography

Abbey, Edward. *Desert Solitaire*. New York: Touchstone Books, 1968.

Abbey, Edward.*The Monkey Wrench Gang*. New York: Harper Perennial, 2006.

Ackroyd, Peter. *London: The Biography*. New York: Anchor Books, 2003.

Adams, Charles Francis. *Railroads: Their Origins Problems*. New York: G. P. Putnam's Sons, 1878.

Adkins, Roy and Lesley Adkins. *Jane Austen's England*. New York: Viking, 2013.

Allaby, Michael. *Fog, Smog, and Poisoned Rain*. New York: Infobase Publishing, 2014.

Allen, Robert C. *The British Industrial Revolution in Global Perspective*. Cambridge: Cambridge University Press, 2009.

Alpers, Paul. *What is Pastoral?* Chicago: University of Chicago Press, 1996.

Anderson, Lorraine and Scott Slovic. *Literature and the Environment: A Reader on Nature and Culture*. 2nd edition. Boston: Pearson Education, 2013.

Archer, John E. *Social Unrest and Popular Protest in England, 1780–1840*. Cambridge: Cambridge University Press, 2000.

Asimov, Isaac. "The Last Question." In *Isaac Asimov: The Complete Stories*, Vol. 1. New York: Doubleday, 1959, p. 290–300.

Attridge, Derek. *The Singularity of Literature*. New York: Routledge, 2004.

Austen, Jane. *Emma*. Fiona Stafford, ed. London: Penguin Classics, 2003.

Austen, Jane. *Mansfield Park*. Ian Littlewood, ed. Ware, UK: Wordsworth Editions, 2000.

Austen, Jane. *Northanger Abbey*. James Kinsley and John Davie, eds. Oxford: Oxford World's Classics, 2003.

Austen, Jane. *Persuasion*. James Kinsley, ed. Oxford: Oxford World's Classics, 2004.

Austen, Jane. *Pride and Prejudice*. James Kinsley, ed. Oxford: Oxford World's Classics, 1998.

Austen, Jane. *Sense and Sensibility*. Edward Copeland, ed. Cambridge: Cambridge University Press, 2006.

Bacigalupi, Paolo. *Ship Breaker*. New York: Little, Brown, 2010.

Bacigalupi, Paolo. *The Windup Girl*. San Francisco: Night Shade Books, 2009.

Bate, Jonathan. *John Clare: A Biography*. London: Farrar, Straus, and Giroux, 2003.

Beggs, Clive. *Energy Management, Supply, and Conservation*. London: Elsevier Press, 2009.

Berger, Albert. "Nuclear Energy: Science Fiction's Metaphor of Power." *Science Fiction Studies*, 6 (1979): 121–7.

Bergthaller, Hans. "Fossil Freedoms: The Politics of Emancipation and the End of Oil." In *The Routledge Companion to the Environmental Humanities*. Ursula Heise and John Christensen, eds. New York: Routledge, 2017, p. 424–32.

Berry, Wendell. "Compromise, Hell!" In *Missing Mountains*. Krisin Johannsen, Bobbie
    Ann Mason, and Mary Ann Taylor-Hall, eds. Nicholasville, KY: Wind Publications,
    2005, p. 155–60.

Berry, Wendell. *The Unsettling of America: Culture and Agriculture*. San Francisco:
    Sierra Club Books, 1977.

Bhatt, Sujata. "Wine from Bordeaux." In *Atomic Ghost*. John Bradley, ed.
    Minneapolis: Coffee House Press, 1995, p. 102–4.

Boswell, James. *The Life of Samuel Johnson*. London: Penguin Classics, 2008.

Bradley, John, ed. *Atomic Ghost*. Minneapolis: Coffee House Press, 1995.

Branch, Michael. "Are You Serious? A Modest Proposal for Environmental Humor." In
    The Oxford Handbook of Ecocriticism. Greg Garrard, ed. Oxford: Oxford University
    Press, 2014, p. 377–90.

Brimblecomb, Peter. *The Big Smoke: A History of Air Pollution in London since Medieval
    Times*. New York: Routledge, 1987.

Bronte, Charlotte. *The Novels of the Sisters Bronte*. London: J. M. Dent, 1905.

Bronte, Charlotte. *Shirley*. Hertfordshire, UK: Wordsworth Editions, 1993.

Bronte, Emily. *Wuthering Heights*. London: Penguin Classics, 1995.

Buell, Frederick. "A Short History of Oil Cultures; or, the Marriage of Catastrophe
    and Exuberance." In *Oil Culture*. Ross Barrett and Daniel Worden, eds.
    Minneapolis: University of Minnesota Press, 2014, p. 69–88.

Callenbach, Ernest. *Ecotopia Emerging*. Berkeley, CA: Banyan Books, 1981.

Carol [Feature Film]. Dir. Todd Haynes. The Weinstein Company, 2015.

Cartmill, Cleve. "Deadline." In *The Golden Age of Science Fiction*. Groff Conklin, ed.
    New York: Bonanza Books, 1980, p. 67–88.

Christianson, Gale E. *Greenhouse: The Two Hundred Year Story of Global Warming*.
    New York: Walker, 1999.

Cipolla, Carlo. *Before the Industrial Revolution: European Society and Economy
    1000–1700*. 3rd edition. New York: Norton, 1993.

Clark, Timothy. *Ecocriticism on the Edge: The Anthropocene as a Threshold Concept*.
    London: Bloomsbury Press, 2015.

Coleridge, Samuel Taylor. *Coleridge's Poetry and Prose*. New York: W. W. Norton, 2004.

Collins, Wilkie. *Miss or Mrs?: The Haunted Hotel, The Guilty River*. Oxford: Oxford
    World's Classics, 1999.

Crace, Jim. *Harvest*. New York: Vintage Books, 2013.

Craik, Dinah. *John Halifax, Gentleman*. Toronto, Ont.: Broadview Editions, 2005.

Dee, Tim. *Four Fields*. Berkeley, CA: Counterpoint Press, 2015.

Defoe, Daniel. *Robinson Crusoe*. New York: W. W. Norton, 1975.

DeQuincey, Thomas. *The Collected Writings of Thomas DeQuincey*, Vol. 13. London: A
    & C Black, 1897.

Diamond, Jared. *Collapse: How Societies Choose to Fail or Succeed*. New York: Penguin, 2005.

Dickens, Charles. *Bleak House*. New York: Charles Scribner's Sons, 1901.

Dickens, Charles. *Dombey and Son*. London: Penguin, 2002.

Dickens, Charles. *Hard Times*. Oxford: Oxford University Press, 2008.

Dr. No. Dir. Terence Young. 1962. Sony Pictures Home Entertainment, 2006. DVD.

"Driving Destruction in the Amazon: How Steel Production Is Throwing the Forest into the Furnace." *Greenpeace International*. 2nd edition. February 2013.

Eberhart, Mark. *Feeding the Fire: The Lost History and Uncertain Future of Mankind's Energy Addiction*. New York: Crown Publishing, 2007.

Edgecombe, Rodney Stenning. *Leigh Hunt and the Poetry of Fancy*. Madison, NJ: Fairleigh Dickinson University Press, 1994.

Eliot, George. *The Mill on the Floss*. Carol Christ, ed. New York: W. W. Norton, 1994.

Engels, Friedrich. *The Condition of the Working Class in England*. Mark Harris, ed. London: Panther Editions, 2010.

"Environmental Benefits of Moving Freight by Rail." Association of American Railroads, April 2014.

Fagan, Brian. *The Little Ice Age: How Climate Made History 1300–1850*. New York: Basic Books, 2001.

Fleming, Ian. *Moonraker*. 1955. London: Coronet, 1989.

Flenniken, Kathleen. *Plume*. Seattle: University of Washington Press, 2012.

Freese, Barbara. *Coal: A Human History*. New York: Penguin Books, 2003.

"Fukushima-Related Measurements by the CTBTO." CTBTO Preparatory Commission. April 13, 2011. http://www.ctbto.org/press-centre/highlights/2011/fukushima-related-measurements-by-the-ctbto/fukushima-related-measurements-by-the-ctbto-page-1/.

Gardner, J. S. and P. Sainato. "Mountaintop Mining and Sustainable Development in Appalachia." *Mining Engineering* (March 2007): 48–55.

Graham, Hamish. "'Alone in the Forest'? Trees, Charcoal, and Charcoal Burners in Eighteenth-Century France." In *Invaluable Trees: Cultures of Nature, 1660–1830*. Laura Auricchio, Elizabeth Hechendorn Cook and Giulia Pacini, eds. Oxford, UK: Voltaire Foundation, 2012, p. 23–37.

Garrard, Greg. "Solar: Apocalypse Not." In *Ian McEwan*. 2nd edition. Sebastian Groes, ed. London: Bloomsbury Press, 2013, p. 123–36.

Gaskell, Elizabeth. *North and South*. New York: Oxford University Press, 1998.

Gill, Jo. *The Poetics of the American Suburbs*. New York: Palgrave Macmillan, 2013.

Ginsberg, Allen. "Plutonium Ode." In *Atomic Ghost*. John Bradley, ed. Minneapolis: Coffee House Press, 1995, p. 245–9.

Gordon, John. "Gaslight, Ghostlight, Golliwog, Gaslight." *James Joyce Quarterly*, vol. 46, no. 1 (Fall 2008): 19–37.

Graham, Hamish. "'Alone in the Forest'? Trees, Charcoal, and Charcoal Burners in Eighteenth-Century France." In *Invaluable Trees: Cultures of Nature, 1660–1830*. Laura Auricchio, Elizabeth Hechendorn Cook, and Giulia Pacini, eds. Oxford, UK: Voltaire Foundation, 2012, p. 23–37.

Greene, Catherine. "Growth Patterns in the U.S. Organic Industry." USDA Economic Research Service. October 24, 2013. http://www.ers.usda.gov/amber-waves/2013-october/growth-patterns-in-the-us-organic-industry.aspx#.UwzdN4Wu4j4.

Grey, Zane. *Boulder Dam*. New York: HarperCollins, 1963.

"Hanford Cleanup." US Department of Energy. http://www.hanford.gov/page.cfm/
    HanfordCleanup.

Hansen, Gary. *Wet Desert*. Hole Shot Press, 2007.

Hardy, Thomas. *Tess of the D'Urbervilles*. Peterborough, Ont.: Broadview Press, 2003.

Haugeland, John. "Reading Brandom Reading Heidegger." *European Journal of
    Philosophy* 13 (2005): 421–8.

Haushofer, Marlen. *The Wall*. Berkeley, CA: Cleis Press, 1990.

Heaney, Seamus. *Opened Ground: Selected Poems 1966–1996*. New York: Farrar, Strauss,
    Giroux, 1998.

Heinlen, Robert. "Blowups Happen." In *The Golden Age of Science Fiction*. Groff
    Conklin, ed. New York: Bonanza Books, 1980, p. 103–39.

Helmholtz, Hermann von. "On the Interaction of Natural Forces." John Tyndall, trans.
    London, Edinburgh, and Dublin Philosophical Magazine and Journal of Science, vol.
    11, no. 4 (1856).

Henderson, Heather and William Sharpe. *The Longman Anthology of British Literature,
    Vol. 2B: The Victorian Age*. New York: Addison-Wesley Educational Publishers, 2003.

Herold, Anke. "ETC/ACC Technical Paper 2003/10." European Topic Centre on Air and
    Climate Change, July 2003.

Highsmith, Patricia. *The Price of Salt*. New York: Dover Publications, 2015.

Hilliard, Jeffrey. "The Message." In *Atomic Ghost*. John Bradley, ed. Minneapolis: Coffee
    House Press, 1995, p. 97–8.

Hiltner, Ken. "Coal in the Age of Milton." *PMLA*, vol. 126, no. 2 (2011): 316–18.

Hitchcock, Peter. "Oil in an American Imaginary." *New Formations*, 69 (2010).

Hoban, Russell. *Riddley Walker*. Bloomington, IN: Indiana University Press, 1980.

"Hoover Dam." Bureau of Reclamation Historic Dams and Water Projects. http://
    www.nps.gov/nr/travel/ReclamationDamsAndWaterProjects/Hoover_Dam.html.
    Accessed June 3, 2014.

Hughes, Merritt Y., ed. *John Milton: Complete Poems and Major Prose*. Upper Saddle
    River, NJ: Prentice-Hall, 1957.

Hurd, Barbara. "Fracking: A Fable." *Brevity*, 42 (March 2013).

Irving, Washington. *The Legend of Sleepy Hollow and Other Tales*. New York: Modern
    Library, 2001.

"Japan Shuts Down Last Nuclear Reactor—for Now." CNN, Sept 16, 2013. http://www.
    cnn.com/2013/09/15/world/asia/japan-nuclear-reactor-shutdown/.

"Japan: Train Fans Experience Super-Fast Maglev Speed." *BBC News*, Nov. 15, 2014.
    http://www.bbc.com/news/blogs-news-from-elsewhere-30051961.

Johnson, Bob. *Carbon Nation*. Lawrence: University Press of Kansas, 2014.

Jordan, June. "Directions for Carrying Explosive Nuclear Wastes through Metropolitan
    New York." In *Atomic Ghost*. John Bradley, ed. Minneapolis: Coffee House Press,
    1995, p. 91–2.

Kelley, D. W. *Charcoal and Charcoal-Burning*. Oxford, UK: Shire Publications, 1986.

Kerouac, Jack. *On the Road*. London: Penguin, 2003.

Ketabgian, Tamara. *The Lives of Machines: The Industrial Imaginary in Victorian Literature and Culture*. Ann Arbor: University of Michigan Press, 2011.

Konstam, Angus. *The Armada Campaign 1588: The Great Enterprise against England*. Oxford: Osprey Publishing, 2001.

Krakauer, John. "Death of an Innocent." *Outside* (January 1993): 40–5, 90–2.

Lamb, Charles. "The Londoner." In *Charles Lamb: Select Writings*. J. E. Morpurgo, ed. New York: Routledge, 2003, p. 162–4.

Laucht, Christoph. "Britannia Rules the Atom: The James Bond Phenomenon and Postwar British Nuclear Culture." *The Journal of Popular Culture*, vol. 46, no. 2 (2013): 358–77.

LeMenager, Stephanie. "The Aesthetics of Petroleum, after *Oil!*" *American Literary History*, vol. 24, no. 1 (2012): 59–86.

LeMenager, Stephanie. *Living Oil*. Oxford: Oxford University Press, 2014.

Ling, Katherine. "Nuclear Power Cannot Solve Climate Change." *Scientific American* (March 27, 2009). http://www.scientificamerican.com/article/nuclear-cannot-solve-climate-change/.

Lipow, Gar. "Flying Energy Generators: Maybe the Next Big Thing." *Grist* (August 3, 2010). Online. http://grist.org/article/flying-energy-generators-a-breakthrough-ready-to-happen/. Accessed June 5, 2014.

Llewellyn, Richard. *How Green Was My Valley*. New York: Simon and Schuster, 2009.

London, Jack. *The Best Short Stories of Jack London*. New York: Doubleday, 1945.

Lyall, Sarah. "Bark Up or Down? Firewood Splits Norwegians." *New York Times*. February 19, 2013.

MacDuffie, Allen. *Victorian Literature, Energy, and the Ecological Imagination*. Cambridge: Cambridge University Press, 2014.

Mad Max: Fury Road [Feature Film]. Dir. George Miller. Warner Brothers Pictures, 2015.

Massey, Nathanael. "Nuclear Power Also Needed to Combat Climate Change." *Scientific American* (April 16, 2014). http://www.scientificamerican.com/article/nuclear-power-also-needed-to-combat-climate-change/.

McEwan, Ian. *Solar*. New York: Random House, 2010.

McInerney, Jeremy. *The Cattle of the Sun: Cows and Culture in the World of the Ancient Greeks*. Princeton: Princeton University Press, 2010.

McKusick, James. *Green Writing: Romanticism and Ecology*. New York: St. Martin's Press, 2000.

McPhee, John. *Coming Into the Country*. New York: Farrar, Strauss, Giroux, 1976.

McPhee, John. *The Pine Barrens*. New York: Farrar, Strauss, Giroux, 1967.

Melville, Herman. *Moby-Dick*. New York: Pearson/Longman, 1997.

Menely, Tobais and Jesse Oak Taylor, eds. *Anthropocene Reading: Literary History in Geologic Times*. University Park, PA: Pennsylvania State University Press, 2017.

Metras, Gary. "Utilities Advertisement in the Wake of Three Mile Island (a Found Poem)." In *Atomic Ghost*. John Bradley, ed. Minneapolis: Coffee House Press, 1995, p. 96.

Moller, Hayley. "Hydropower Continues Steady Growth." Earth Policy Institute. June
    14, 2012. http://www.earth-policy.org/data_highlights/2012/highlights29. Accessed
    June 12, 2014.

Moomaw, W. et al., 2011: Annex II: Methodology. In IPCC Special Report on
    Renewable Energy Sources and Climate Change Mitigation. O. Edenhofer, R. Pichs-
    Madruga, Y. Sokona, K. Seyboth, P. Matschoss, S. Kadner, T. Zwickel, P. Eickemeier,
    G. Hansen, S. Schlomer, C. von Stechow, eds., Cambridge University Press,
    Cambridge, UK and New York.

Morgan, Ted. *Wilderness at Dawn: The Settling of the North American Continent.* New
    York: Simon and Schuster, 1993.

Morin, G. A. et al. *Long-Term Historical Changes in the Forest Resource.* New York:
    United Nations, 1996.

Morton, Timothy. *Hyperobjects: Philosophy and Ecology after the End of the World.*
    Minneapolis: University of Minnesota Press, 2013.

Morton, Timothy. "Queer Ecology." *PMLA*, vol. 125, no. 2 (2010): 273–82.

Muir, John. *John Muir: Nature Writings.* William Cronon, ed. New York: Library of
    America, 1997.

Nagourney, Adam and Ian Lovett. "Severe Drought has U.S. West Fearing Worst."
    New York Times, February 1, 2014.

Nardizzi, Vin. "Wooden Slavery." *PMLA*, vol. 126, no. 2 (2011): 313–5.

Nead, Lynda. *Victorian Babylon.* New Haven, CT: Yale University Press, 2000.

Nickerson, Billeh. *McPoems.* Vancouver: Arsenal Pulp Press, 2009.

Nixon, Rob. *Slow Violence and the Environmentalism of the Poor.* Cambridge: Harvard
    University Press, 2011.

Nye, David E. *Consuming Power.* Cambridge, MA: MIT Press, 1998.

O'Neill, Eugene. *Three Plays.* New York: Boni and Liveright, 1922.

Otis, Laura. *Literature and Science in the Nineteenth Century.* Oxford: Oxford World's
    Classics, 2002.

Padgett, Lewis. "The Piper's Son." In *The Golden Age of Science Fiction.* Groff Conklin,
    ed. New York: Bonanza Books, 1980, p. 45–66.

Pancake, Ann. *Strange As This Weather Has Been.* Berkeley, CA: Counterpoint Press, 2007.

Pandora's Promise [Documentary Film]. Dir. Robert Stone. Robert Stone Productions,
    2013.

Pater, Walter. *The Renaissance.* New York: Modern Library, 1873.

Penderey, David. The machine at Marly. marlymachine.org.

Picard, Liza. *Victorian London.* New York: St. Martin's Press, 2005.

Pollan, Michael. *The Omnivore's Dilemma.* New York: Penguin Books, 2006.

"Post-Chernobyl Disaster Sheep Controls Lifted on Last UK farms." *BBC News.* June 1,
    2012. http://www.bbc.com/news/uk-england-cumbria-18299228.

Revill, George. *Railway.* London: Reaktion Books, 2012.

Rochan, M. "Chinese New Year 2014: China Urges Firework Ban as it Grapples Smog
    Crisis." *International Business Times*, Jan 30, 2014.

Rothenberg, Jerome and Robinson, Jeffrey C. *Poems for the Millennium*. Berkeley: University of California Press, 2009.

Ruskin, John. *The Stones of Venice*. London: Smith, Elder, 1851.

Ruskin, John. *The Works of John Ruskin*, Vol. 3. New York: Longmans, Green, 1903.

Ruskin, John. *The Works of John Ruskin*, Vol. 16. New York: Longmans, Green, 1905.

Sewell, Anna. *Black Beauty*. New York: Dodge Publishing, 1907.

Shannon, Laurie. "Greasy Citizens and Tallow-Catches." *PMLA*, vol. 126, no. 2 (2011): 311–13.

Shelley, Percy. "A Defence of Poetry." In *The Works of Percy Bysshe Shelley in Verse and Prose, Now First Brought Together with Many Pieces Not Before Published*, Vol. 7. Harry Buxton Forman, ed. London: Reeves and Turner, 1880.

Shimomura, Kazuko. "Let's Listen to the Voiceless Voice." In *Reverberations from Fukushima: 50 Japanese Poets Speak Out*. Leah Stenson and Asao Sarukawa Aroldi, eds. Portland, OR: Inkwater Press, 2014.

Simak, Clifford. "Lobby." In *The Golden Age of Science Fiction*. Groff Conklin, ed. New York: Bonanza Books, 1980, p. 89–102.

Sinclair, Upton. *Oil!* London: Penguin Books, 2007.

Sirota, Lyubov. "Radiophobia." Leonid Levin and Elisavietta Ritchie, trans. In *Atomic Ghost*. John Bradley, ed. Minneapolis: Coffee House Press, 1995, p. 106–7.

Smil, Vaclav. *Energy in World History*. Westview Press, 1994.

Spiegel, Maura, ed. *The Jungle* by Upton Sinclair. New York: Barnes and Noble Classics, 2003.

Steinman , George. *A History of Croydon*. London: Longman, Rees, Orme, Brown, Green, & Longman, 1834.

Stenson, Leah and Asao Sarukawa Arnoldi, eds. *Reverberations from Fukushima: 50 Japanese Poets Speak Out*. Portland, OR: Inkwater Press, 2014.

Stevenson, Robert Louis. *The Strange Case of Dr. Jekyll and Mr. Hyde*. Peterborough, UK: Broadview Press, 2005.

Stevenson, W. H. *Blake: The Complete Poems*. New York: Routledge, 2014.

Stoker, Bram. *Dracula*. London: Penguin Books, 2003.

Swift, Jonathan. *Gulliver's Travels*. New York: Dell Publishing, 1971.

Szeman, Imre. "Literature and Energy Futures." *PMLA*, vol. 126, no. 2 (2011): 323–5.

Tarr, Joel A. "Urban Pollution—Many Long Years Ago." *American Heritage Magazine*, vol. 22, no. 6 (October 1971).

Taylor, Bayard. *Views Afoot*. London: Samson, Low, Son, and Morrison, 1869.

Taylor, Jesse Oak. *The Sky of Our Manufacture: The London Fog in British Fiction from Dickens to Woolf*. Charlottesville: University of Virginia Press, 2016.

Texas Energy Museum displays—Beaumont, TX. Researched May 2013.

"Texas: King of the Wildcatters," *Time*, February 13, 1950, 18–21.

Turpin, Trevor. *Dam*. London: Reaktion Books, 2008.

Ure, Andrew. *The Philosophy of Manufacturers*. London: Charles Knight, 1835.

Velten, Hannah. *Beastly London*. London: Reaktion, 2013.

Virgil. *The Georgics*. L. P. Wilkinson, trans. New York: Penguin Books, 1982.

Wan, William. "Desperate for Clean Air, Chinese Get Creative." *Washington Post.* January 26, 2014.

Weisman, Alan. *The World without Us.* New York: St. Martin's Press, 2007.

Werrett, Simon. "The Power of Pyrotechnics." *History Today* (November 2010): 10–16.

Westmacott, Molloy. *The English Spy.* London: Sherwood, Gilbert and Piper, 1826.

Wilde, Oscar. *The Picture of Dorian Gray.* London: Simpkin, Marshall, Hamilton, Kent, 1891.

Williamson, Tom. "The Management of Trees and Woods in Eighteenth Century England." In *Invaluable Trees: Cultures of Nature, 1660–1830.* Laura Auricchio, Elizabeth Hechendorn Cook, and Giulia Pacini, eds. Oxford: Voltaire Foundation, 2012, p. 221–35.

Wilson, Alexander. *The Culture of Nature.* Cambridge, MA: Blackwell Press, 1992.

Wilson, E. O. *Biophilia.* Cambridge, MA: Harvard University Press, 1986.

Wines, Michael. "Colorado River Drought Forces a Painful Reckoning for States." *New York Times.* January 5, 2014.

Witchard, Anne. "'A Fatal Freshness': Mid-Victorian Suburbophobia." In *London Gothic.* Lawrence Phillip and Anne Witchard, eds. London: Continuum Press, 2010, p. 23–40.

Wolfe, Jessica. *Humanism, Machinery, and Modern Warfare.* Cambridge: Cambridge University Press, 2004.

Wolmar, Christian. *Fire and Steam: A New History of the Railways in Britain.* London: Atlantic Books, 2007.

Wrangham, Richard. *Catching Fire.* New York: Basic Books, 2009.

Wrigley, E. A. *Energy in the English Industrial Revolution.* Cambridge: Cambridge University Press, 2010.

Yaeger, Patricia. "Editor's Column: Literature in the Ages of Wood, Tallow, Coal, Whale Oil, Gasoline, Atomic Power, and Other Energy Sources." *PMLA*, vol. 126, no. 2 (2011): 305–26.

Yonge, Charlotte. *The Daisy Chain.* New York: McMillan, 1876.

Young, Kevin. *The Hungry Ear: Poems of Food and Drink.* New York: Bloomsbury, 2012.

Ziser, Michael. "Oil Spills." *PMLA*, vol. 126, no. 2 (2011): 321–3.

Index